Cambridge Elements ≡

Elements in The Philosophy of Biology
edited by
Grant Ramsey
KU Leuven, Belgium
Michael Ruse
Florida State University

UNITS OF SELECTION

Javier Suárez
Jagiellonian University and University of Oviedo

Elisabeth A. Lloyd
Indiana University

CAMBRIDGE
UNIVERSITY PRESS

Shaftesbury Road, Cambridge CB2 8EA, United Kingdom

One Liberty Plaza, 20th Floor, New York, NY 10006, USA

477 Williamstown Road, Port Melbourne, VIC 3207, Australia

314–321, 3rd Floor, Plot 3, Splendor Forum, Jasola District Centre,
New Delhi – 110025, India

103 Penang Road, #05–06/07, Visioncrest Commercial, Singapore 238467

Cambridge University Press is part of Cambridge University Press & Assessment,
a department of the University of Cambridge.

We share the University's mission to contribute to society through the pursuit of
education, learning and research at the highest international levels of excellence.

www.cambridge.org
Information on this title: www.cambridge.org/9781009449236

DOI: 10.1017/9781009276429

First published 2023

A catalogue record for this publication is available from the British Library

ISBN 978-1-009-44923-6 Hardback
ISBN 978-1-009-27641-2 Paperback
ISSN 2515-1126 (online)
ISSN 2515-1118 (print)

Units of Selection

Elements in the Philosophy of Biology

DOI: 10.1017/9781009276429
First published online: August 2023

Javier Suárez
Jagiellonian University and University of Oviedo

Elisabeth A. Lloyd
Indiana University

Author for correspondence: Javier Suárez, javier.suarez@uniovi.es

Abstract: This Element introduces the disambiguating project (DP) about the units of selection. By DP, the authors mean the thesis that the expression 'units of selection' refers to at least three non-co-extensional functional concepts: interactor, replicator/reproducer/reconstitutor, and manifestor of adaptation/type-1 agent. They present each concept and demonstrate the necessity of their isolation, because each of them responds to a distinct question about the units of selection, and these distinct questions are not always posed in combination in today's biological research. They further apply the framework to the analysis of the debates concerning the evolutionary transitions in individuality (ETI) and argue that the DP interprets the ETI better than any project rejecting the three meanings of 'units of selection'. Thus, they claim that the differentiation between at least these three functional concepts is fundamental to clarify some conceptual confusions in biology, which rest on the conflation of these distinct meanings.

Keywords: levels of selection, evolution by natural selection, Darwinian individuality, evolutionary transitions in individuality, kin selection, multilevel selection, adaptationism

ISBNs: 9781009449236 (HB), 9781009276412 (PB), 9781009276429 (OC)
ISSNs: 2515-1126 (online), 2515-1118 (print)

Contents

Introduction   1

1 What Is a Unit of Selection and How Can We Identify It?
The Disambiguating and the Unitary Projects   3

2 How the Expression 'Units of Selection' Acquired Its
Polysemic Meaning or Why the Disambiguating Project
Started   10

3 Two Sources of Misunderstanding in Past and Today's
Debates about Units   28

4 The Framework of the Evolutionary Transitions in
Individuality: A Challenge to the Disambiguating Project   41

Conclusion   70

Glossary   74

References   76

Introduction

This Element introduces what we call the disambiguating project (DP) about the units or levels of selection. By the DP we mean the thesis that the expression 'units of selection' refers to at least three different non-co-extensional functional concepts: interactor, replicator/reproducer/reconstitutor, and manifestor of adaptation/type-1 agent. We present each of these concepts and demonstrate the necessity of their isolation because each of them responds to a distinct question about the units of selection, and these distinct questions are not always posed in combination in today's biological research. We further apply our framework to the study of the debates concerning the evolutionary transitions in individuality (ETI) and argue that the DP interprets the ETI better than any project rejecting the polysemy (multiple meanings) of the expression. Thus, we claim that the differentiation between at least three functional concepts equivocally meant by the expression 'units of selection' is fundamental to clarify some conceptual confusions in biology, which we argue rest on the conflation of these distinct meanings.

Our project is partially presented as a response to some criticisms by some recent approaches to the study of units that treat the expression as unambiguous. We refer to the latter as the unitary project (UP), a project that has its origin in Lewontin's formulation of the problem of the units of selection in 1970, but that has substantially evolved afterwards. While authors working under the UP find that the expression 'units of selection' may be confusing and requires philosophical treatment, they reject that it stands for more than one concept. Under the UP, disagreements about the units may rest on confusions over the application of *the* concept to different cases, or on empirical disagreements, but are not the result of a conceptual conflation of different meanings under the single expression 'units of selection'. We demonstrate that adopting a UP is not a profitable research avenue to treat the debates about the units of selection, and acknowledging that the expression stands for different functional concepts is essential for proper communication among today's scholars.

Our agenda is as follows: we first introduce our proposal and review some of the historical origins of the DP, and its conceptual relationship to Lewontin's original version of UP, the so-called recipe approach. We argue that the DP does not *oppose* the recipe approach, but rather specifies and connects it to several lines of biological research. We demonstrate the value of using the DP, rather than the UP, by demonstrating that recent evidence from biology and philosophy shows the continuing necessity of distinguishing different non-co-extensional meanings of the expression 'units of selection' to clarify today's debates in biology. More specifically, we show that in today's biological research – especially when properly framed under the adaptationist versus the evolutionary change school of evolution – the expression 'units of selection' refers to at least three different types of distinct functional concepts: interactors, reproducers/replicators, or manifestors of adaptation/type-1 agents. We isolate and analyze these concepts, and argue that each of them singles out a set of independent research questions that can be asked, *either singly or in combination*, about the action of natural selection.

We argue further that the failure to appreciate the existence of the multiplicity of meanings and concepts that we advocate in this work derives from a series of *misconceptions about the DP*. All of these misconceptions are ultimately grounded on an excessive and misguided emphasis on reducing all questions about units to questions framed only under the single adaptationist and reproduction-centred project of the ETI. We contend that this emphasis has caused the emergence of an eliminative version of the UP, which we argue is not a profitable research avenue, neglecting a substantial amount of contemporary biological research about units not directly concerned with the origin of the reproductive hierarchy. We show that the changes involved in the ETI can be straightforwardly analyzed under our version of the DP, and in a more illuminating manner than they are thought of when they are conceived under the tenets of a UP. We thus conclude that the DP is the most adequate way of addressing today's philosophical debates about units.

Highlights

- We distinguish two types of projects about the units of selection: the DP and the UP.
- The DP seeks to clarify different non-co-extensional meanings of the expression 'units of selection' that are used in different research projects.
- The UP conceives that the debate about units only encompasses one type of unit, and its purpose is finding criteria to determine whether an object is a unit of selection.
- We review the DP in a historic-systematic way and demonstrate a tripartite version of it as the most profitable way of characterizing the units of selection debates.
- We relate the DP to Lewontin's recipe approach to the levels of selection and show that the former is a specification of the latter.
- We argue that the criticisms of the DP are ungrounded and based on misunderstanding of the framework.
- We argue that the tripartite version of the DP that we defend can be used to gain a better understanding of the debates about the ETI.

1 What Is a Unit of Selection and How Can We Identify It? The Disambiguating and the Unitary Projects

The units or levels-of-selection debate concerns the type of biological forms of organization that can evolve by means of the process of natural selection, originally described by Charles Darwin in *The Origin of Species*. According to Darwin's process, under ideal conditions, the traits of a biological form of organization (e.g., organisms and colonies) that are systematically linked to the fitness of their bearers and confer on them a relative fitness advantage with relation to other members of their population will spread in the population until they become the predominant forms (Box 1).

Darwin's process is additionally the basic schema for explaining the existence of highly sophisticated traits that make some organisms – or alleles, groups, or colonies – look as if they were engineered to fit their environment (engineering/trans-temporally accumulated adaptations). For example, it is used to explain why anteaters have their characteristic elongated noses and tongues way longer than their heads and lips, but no teeth in their mouths. Or why vampire bats have their teeth, sensory apparatus, kidney, and bladder adapted for a purely blood-sucking diet. Or how collared flycatchers have their characteristic coloured patterns, especially their spectacular white collar. The explanation of why all these organisms and their populations possess these phenotypic characteristics lies in a basic idea: these spectacular traits are connected to the relative fitness of these organisms and their populations, such that anteaters with the aforementioned characteristics reproduced more than those that lacked them, until a point where the traits become the only extant forms in the population of anteaters. The same would be true for the traits of vampire bats and flycatchers.

An intriguing philosophical question about the units of selection concerns *the ways of singling out* the biological forms of organization that have the required properties to be causally affected by natural selection and/or to evolve by natural selection. This fundamental philosophical question cannot be divorced from the *biological question about what natural selection is*, for the way in which one replies to the latter is intimately connected to how one replies to the former, as we will show. So, the question about the units of selection is in the end simultaneously a question about how the process that Darwin described in *The Origin of Species* should be understood. In other words, we have to know *what the process really is*, before we can understand the *best ways to look for it*.

A long tradition in philosophy of biology that dates back to the early works of Dawkins (1976) and Hull (1980), and continues with the works of Wimsatt (1980a, 1980b); Brandon (1981); or Lloyd (1986, 1988/1994, 1992, 2001, 2023; Lloyd and Wade 2019) conceives that the philosophical question about the units

Box 1 Darwin's process and the evolutionary process are not the same

It is sometimes mistakenly assumed that Darwin's process (or evolution by natural selection) is equivalent to the evolutionary process, or, to put it differently, that evolution is made up only of evolution by natural selection. This is however a misconception of the aims of contemporary evolutionary research. Contemporary evolutionary research is concerned with the study of several processes and mechanisms (or, more generally, *evolutionary factors*) that change the composition of biological populations over time differently, sometimes in combination favouring the same type of changes, but others with some processes opposing the direction of others.

Some of the evolutionary mechanisms, like genetic mutations, recombination, or genetic drift, affect the pattern of inheritance, and in doing so affect the evolutionary process. Others, like niche construction, phenotypic plasticity, or other developmental processes, affect the patterns of expression of traits, and in doing so they also affect the evolutionary process. Finally, a third group of processes includes phenomena like phyletic inertia, evolutionary trade-offs, or some developmental constraints (see Keller & Lloyd 1992, for discussion of these concepts), which affect the composition of biological populations by opposing certain types of changes due to the intrinsic configuration of the entities composing the biological population.

Under this perspective, the aim of evolutionary biology is to explain the history, relatedness, and the forms and function of life on Earth (e.g., Hull 1988a; Gould 2002; Lloyd 2015, 2021), but without necessarily highlighting the influence of one mechanism – natural selection – or one specific output – engineering/trans-temporally accumulated adaptations – over the rest. In this complex view of the evolutionary process, Darwin's process of natural selection is a way of affecting the evolution of a population by favouring the trait variants that increase the fitness of their bearers. But Darwin's process alone does not *solely determine* the composition of biological populations, or how they change over time, because the type of variants it favours may interact with or be opposed by any other evolutionary process. This Element is only dedicated to the study of the Darwinian process, so our analysis may rest on some idealizations which we do not want the reader to misunderstand.

of selection invites a project for distinguishing or disambiguating different meanings of the expression. Under this project, it is assumed that biologists and philosophers sometimes disagree about what the units of selection are because they mean different concepts by the same expression. Thus, part of the disputes about the units may fade away as soon as one realizes that some researchers are pursuing different legitimate questions, each of which triggers a different debate.

The task of a philosopher working under this type of project would consist in disambiguating the different meanings of the expression – that is, they would clarify the polysemic character that 'unit of selection' can adopt in different research contexts. A central guiding assumption of some researchers working under this tradition is the acknowledgement of the existence of at least two different approaches to evolution – later characterized as the distinct Adaptationist and Evolutionary Change Schools – each guided by a different set of research problems (e.g., Wimsatt 1980a, Griesemer 2000a, 2005; Wade, 1978, 2016; Lloyd 1988/1994, 2023). We refer to the project carried under this prior long tradition as the DP (Box 2).

In contrast, researchers from the adaptationist tradition of the study of ETI or Major Transitions in Individuality have recently denied the polysemy of the expression; they argue that the meaning of 'unit of selection' is unambiguous. They believe there are *not* several philosophical questions about the units, *but a single one* (Okasha 2006; Godfrey-Smith 2009, 2013, 2015).[1] Researchers working under this second project accept that there can be ambiguities or disagreements about *how best to characterize the unit of selection*. But they disagree with those working under the DP that these disagreements rest on a conflation of different concepts or questions under the expression 'units of selection'. In their view, the question is simply *singular*, and the debate is about *one* type of unit whose properties must be discovered. We refer to this tradition as the *unitary project*. Such a view has recently been expressed by Godfrey-Smith, who claims:

> Questions about the "unit" of selection *are not ambiguous*; the units in a selection process are just the entities that make up a Darwinian population at that level. (Godfrey-Smith 2009, p. 111, emphasis added)

The task of a philosopher working under the UP tradition would be to uncover the set of properties that are necessary and sufficient to argue that the entities at one specific level are units of selection (Box 3). We will later show that a peculiarity of researchers working under this tradition is that they neglect the existence of different schools of evolution, each guided by a different research programme aimed at replying to different research questions.

Researchers working under the UP have posed several criticisms to the usefulness and necessity of the DP for today's biological research, some of

---

[1] But see Griesemer's analysis of ETIs (2000a, 2000b, 2005) for exceptions to this tendency to reduce the questions about units to a single problem.

Box 2 A BRIEF INTRODUCTION TO THE TYPES OF UNITS THAT WE DISTINGUISH IN THIS
WORK

We distinguish three functional meanings of the expression 'units of selection', each of which would capture a special set of research questions that can be asked either singly or in combination. These meanings are:

(a) Interactor: Units that interact with the environment in such a way that replication or reproduction is differential. The interactor captures the trait–environment relationship and its effects over the differential fitness of bearers of the trait.

For example, the members of the moth species *Biston betularia* would act as interactors in the classic selection processes such as industrial melanism, in which changes in environmental pollution triggered a fitness advantage for darker moths. Note that while *B. betularia* is an interactor with respect to its colour, it simultaneously has a 'product-of-selection adaptation', since no *new* mechanisms or properties are evolved in, that is, introduced into, the selection process producing the increased frequency of dark moths in the population. The only thing that has changed as a result of the selection process is the *frequency* in the population of dark moths; their biology is unchanged.

(b) Manifestor of adaptation/type-1 agent: A manifestor is a unit where a selection process has acted/acts consistently over time resulting in the accretion of a new mechanism or new process not seen before in the lineage, that is, in a tinkering/engineering or trans-temporally accumulated adaptation. A type-1 agent is a subclass of the manifestor in which the optimization of several traits at the level seems obvious, and where the history and accumulation of selection is responsible for such optimization.

For instance, bee colonies are manifestors of adaptation with respect to the bearded sting of individual bees. In this case, the colony is the manifestor of adaptation because the sting shows optimization at the colony level insofar as its use causes the death of individual bees while simultaneously protecting the colony. On the other hand, humans function as manifestors of adaptation *and* type-1 agents with respect to their eyes, as the eye shows a clear history of optimization and trans-temporal accumulation of multiple traits across phylogenetic history.

(c) Replicator/reproducer/reconstitutor: Unit that gets differentially copied (replicator), differentially transmitted through material overlap (reproducer), or differentially recreated in the absence of copy or material overlap (reconstitutor) across generations. This type of unit is responsible for the process of heritability. A unit playing this functional role must be introduced in the analysis of units to clearly distinguish cases of ontogenesis where changes are due to phenotypic plasticity, from cases where the changes can be due to the action of natural selection (see Keller & Lloyd 1992 for discussion of these concepts).

For example, genes are replicators for traits such as eye colour; gametes or whole cells are reproducers for epigenetic traits such as certain disease susceptibilities in humans; many holobionts (animal or plant hosts plus their microbiome), such as vampire bats, are reconstitutors for traits like sanguivory.

Note that this is only a brief introduction, as the aim of this Element is to make these meanings more precise, while tracing back their historical development and the reasons why they must be kept in mind when analyzing units.

which have not yet received a systematic response. This Element presents a tripartite and functionalist version of the DP against the body of these criticisms.[2] Our tripartite version rearticulates and makes clearer some of the meanings of the expression 'units of selection' originally isolated by a variety of researchers working in the DP tradition. Concretely, we isolate three meanings of the expression 'units of selection' and connect each of them to (i) a specific research question, (ii) the type of evidence required to reply to the research question affirmatively, (iii) the type of modelling practices used to reply to the question, and (iv) the predominant research context where the question is usually asked. Grounded on this, we also show that some of these questions, while conceptually distinct, are asked in combination in research on the ETI.

This Element argues for the necessity of the tripartite version of the DP to solve some of the more pressing debates in today's biology. We focus on debates where biologists disagree because they are using different meanings of the expression 'units of selection'.

Our central message is that the historical proliferation of different versions of the DP since its introduction in the 1980s responds to the existence of different types of research questions in biological research about units, each motivated by different ways of perceiving how the processes of natural selection and evolutionary change ultimately act (Box 1). This variety of factors and processes includes the type of units these processes act upon, the type of outcomes they may produce, and the type of evidence that needs to be gathered to demonstrate

---

[2] By *functionalist* we mean that each of the meanings we isolate is distinguished by the specific causal role it plays in the process of natural selection.

Box 3 UNIT OF SELECTION VERSUS LEVEL OF SELECTION VERSUS TARGET OF SELECTION

The expressions 'units of selection' and 'levels of selection' are often used interchangeably (Okasha 2006; Godfrey-Smith 2009; Bourrat 2021; Lloyd 2023), but occasionally they have been used to refer to different concepts (Brandon 1988). As a matter of fact, in biological debates the two expressions are not usually distinguished, being used synonymously by most authors, and sometimes ambiguously with respect to the specific functional meaning intended. As this Element is addressed to these biologists, as well as philosophers of biology interested in these debates, all of which tend not to distinguish the expressions, we will use them interchangeably here.

Metaphysically speaking, however, there are principled reasons to differentiate between them. 'Unit of selection' generally refers to a functional meaning, be it interactor, manifestor/type-1 agent, or replicator/reproducer/reconstitutor, which can apply across different levels of the biological hierarchy. 'Level of selection' and 'target of selection', on the other hand, may be used to refer to a formal role, or a specific level of entity in the biological hierarchy, for example, gene, genome, cell, organism, group, and colony. In this sense, 'unit of selection' would be more general/abstract, and would be the genuine object of philosophical inquiry: how many [functional, formal, or structural] meanings of 'unit' are there? What are the abstract, structural/formal, or phenomenal criteria to distinguish types of units from one another? Are these principled criteria or do some meanings reduce to other meanings? All these constitute genuine philosophical questions about the units.

'Levels of selection' and 'targets of selection', in contrast, could be reserved for the empirical study of the biological objects that may function as units of selection, in any of its meanings: are bee colonies interactors/targets of selection? Are bee colonies also manifestors/type-one agents? Are holobionts reproducers as a whole? Or only partly? All these constitute empirical questions about specific objects in the biological hierarchy.

that selection and/or evolutionary change is acting at that level. In today's biology, we postulate that there are *at least* three, perhaps four or more, non-co-extensional meanings of the expression 'units of selection', each capturing a distinct functional, causal, role that different entities might play in the process of natural selection.

Because we think that the rationale of a research project is not separable from its history, and the type of problems it was aimed to solve when the project originally started, our argument will be presented historically, roughly in chronological order. However, our main claims will be complemented with evidence directly taken from today's biological research along the way.

In Section 2, we introduce the historical origin of the debate(s) about units of selection and justify the reasons why the DP started. We illustrate how the DP evolved, and how at least three different meanings were isolated.

In Section 3, we systematically introduce the two main sources of misunderstanding in debates about units that we have identified. Based on this, we show evidence of how the meanings we isolated in Section 2, and their related research questions – originally isolated by Lloyd – have been reintroduced under different names by biologists and philosophers in recent years. This reintroduction responds to the perception, by both philosophers and biologists analyzing recent debates about units of selection that it is necessary, to distinguish between different types of research questions at stake in these debates. We take this as evidence that the DP is still necessary, both to avoid misunderstandings and to prevent biologists talking past each other.

In Section 4, we show how the Adaptationist version of the ETI project, which started in the 1990s, wrongly convinced some people that the DP was not necessary anymore, leading to the emergence of a series of UPs. We show, in detail, that those who became convinced that the DP was no longer necessary misconceived some fundamental aspects of the project. More importantly, they wrongly conceived *a (specific) project about the evolution of reproduction* – which is what the adaptationist version of ETI is – as if it encompassed *all the projects about units*. We first offer a reinterpretation of this ETI project under the lens of the DP. We then show that the DP is better suited to capture the complexity of the ETI better than any of the currently extant UPs attempting to frame the ETI project. Finally, we show how even some of those who initially rejected the DP have recently re-introduced some of the concepts originally isolated under the same or different names.

Finally, in Section 5, we conclude that the DP is here to stay. Today's evidence still suggests that the expression 'unit of selection' has at least three meanings, referring to three distinct functional roles that trigger three different types of research questions: interactor, replicator/reproducer/ reconstitutor, and manifestor/type-1 agent questions. These different concepts are investigated by different biologists across different schools of evolution. We leave as an open question whether further meanings could be discovered.

2 How the Expression 'Units of Selection' Acquired Its Polysemic Meaning or Why the Disambiguating Project Started

This section introduces the DP in a historic-systematic way. Firstly, we introduce Lewontin's recipe approach as the source of contemporary units of selection debates. We explain the historical and biological reasons why Dawkins and Hull soon perceived the necessity of distinguishing two questions in debates about the units, instead of one as Lewontin's recipe approach implied. We show that Dawkins' and Hull's accounts are not in opposition to the recipe approach, but are a way of specifying its nature. Secondly, we introduce the tripartite Framework in the DP as a way of specifying a further non-co-extensional meaning of the expression 'units of selection' to account for part of the disagreements between researchers working on the evolutionary change school and those working in the adaptationist school of evolution (Section 3). We later show the relationship between the tripartite Framework and the recipe approach, demonstrating that they are compatible and, in fact, complementary approaches to think about units.

2.1 From the Recipe Approach to the Interactor/Replicator Framework

Lewontin (1970) constitutes the classical source for introducing the question about the units of selection as an urgent issue to be resolved in the biological and philosophical agendas. He introduced the debate as follows:

> The principle of natural selection as the motive force for evolution was framed by Darwin in terms of a "struggle for existence" on the part of organisms living in a finite and risky environment. The logical skeleton of his argument, however, turns out to be a powerful predictive system for changes at all levels of biological organization. As seen by present day evolutionists, Darwin's scheme embodies three principles . . .:
>
> 1. Different individuals in a population have different morphologies, physiologies, and behaviours (phenotypic variation).
> 2. Different phenotypes have different rates of survival and reproduction in different environments (differential fitness).
> 3. There is a correlation between parents and offspring in the contribution of each to future generations (fitness is heritable).
>
> These three principles embody the principle of evolution by natural selection. While they hold, a population will undergo evolutionary change. (Lewontin 1970, p. 1)

Lewontin's conception of the units has been referred to as the 'recipe approach' (Okasha 2006). According to it, identifying a level or unit that is subject to the process of natural selection requires identifying objects that possess three characteristics: (1) phenotypic differences, (2) differences in the relative fitness of each phenotypic variant, and (3) heritability of the variance.[3]

Under Lewontin's formulation of natural selection, it was required that the three properties were simultaneously possessed by one single entity in a population of alike entities, and thus the expression 'the units of selection' was univocally connected to the object(s) simultaneously bearing these three properties. That is, something possessing all three qualities could be a 'unit of selection'. This was so even though natural selection could apply hierarchically to different objects (genes, cells, gametes, multicellular organisms, demes, populations, or ecosystems), a fact that Lewontin had empirically demonstrated (Lewontin and Dunn 1960; Lewontin 1962). Yet the characteristic that makes it true that all these different objects at different levels qualify as units of selection, under the later interpretation of Lewontin's criteria that we will criticize in this work, is that every object at each of these levels *simultaneously* possess the three characteristics.

By the time Lewontin published his work, however, it was generally presumed that the Recipe would not work for groups owing to the absence of significant group heritability of mean fitness. This type of scepticism towards the effectiveness of group selection had been and still is near-dogma among many biologists since Williams' (1966), despite Lewontin's empirical and theoretical contributions. For example, Holsinger (1994, p. 626, emphasis added) writes: 'The apparent lack of a mechanism for inheritance for group-level characteristics is the primary reason *gene selectionists* have denied that groups can be units of replicator selection.'

Despite scepticism, the case for group selection as described in the recipe approach was empirically demonstrated in the 1970s through Wade's (1977, 1979) experimental approach to group selection using the flour beetle, *Tribolium castaneum*, in laboratory metapopulations. With himself as the cause of different rates of population survival and reproduction through his laboratory selection, Wade demonstrated that artificial group selection could both increase or decrease mean population size. He also showed that random genetic drift created significant among-population heritability of mean fitness and that this component of heritability increased in the absence of group selection. In a series of papers with McCauley (Wade & McCauley 1980, 1984, 1988), they showed that metapopulations with many different

---

[3] While Lewontin's original formulation requires that heritability affects the variance in *relative fitness*, he later changes this requirement for the notion that what needs to be heritable are instead the very *phenotypic differences* (Levins & Lewontin 1985). We assume that interpretation here. See also Griesemer (2000a) and Godfrey-Smith (2009) for a similar interpretation.

combinations of effective population size and migration rates resulted in significant 'population heritability'. They also developed several different methods for estimating the heritability of population mean fitness, proving the applicability of the recipe approach across different hierarchical levels (see also Bijma et al. 2007a, 2007b; Griesemer & Wade 2000; Wade & Griesemer 1988).

Lewontin's requirement of the three characteristics posed some problems, however, insofar as it did not allow for distinguishing the (potentially) short-term properties that allow the interaction between the organism and its environment and determine its relative fitness (phenotype), from the long-term properties that allow us to trace persistence of the bases of this interaction (genotype). Distinguishing these properties was crucial for understanding the genetic models in the 1960s and 1970s, and hence the characteristics isolated by Lewontin were soon split by Dawkins (1976/2016) and Hull (1980). This resulted in the isolation of two distinct functional properties that do not need to be simultaneously possessed by one single object, but can be satisfied by different types of object; specifically, the *vehicle* (*interactor*, to Hull), which interacts with its environment, and the *replicator*, which rides inside the vehicle and passes on its traits to the future vehicles.

The introduction of these two meanings started the transformation of *the question/debate* about the units of selection into an independent but related *set of research questions/debates*. Or, to put it differently, it caused the debate about units of selection to transform from a UP where determining the defining properties of a bona fide unit of selection was the central research goal, along with identifying genuine cases of such units, into a DP, in which distinguishing distinct types of properties, and hence, different non-co-extensional meanings of the expression 'units of selection', became a central task.

Understanding the confusions that motivated the introduction and hence rationale of the DP, requires going back to the 1960s. By then, a serious theoretical dispute about the biological feasibility of species and group – or more specifically, population – selection had emerged.

In 1962, Wynne-Edwards published his book *Animal Dispersion in Relation to Social Behaviour*, where he postulated that many of the observed and otherwise mysterious animal behaviours actually result from traits selected at the population, *not the organismic*, level. He refers to these traits as 'epideictic' mechanisms or displays. Epideictic mechanisms include, for example, communal nuptial displays, the existence of specific areas for courtship ('leks'), or the formation of aggregations during estivation would be epideictic mechanisms (see Borrello 2010). Wynne-Edwards argues that epideictic mechanisms constitute metapopulational engineering or trans-temporally accumulated adaptations at the species or population level.

The idea of postulating the existence of epideictic mechanisms came to Wynne-Edwards when analyzing previous work about the role of food as a limiting resource for population growth. Drawing on it, he reasoned that if food availability were the only constraint imposed on populations, then competition among individuals for resources would necessarily lead to an overexploitation of the environment, and hence to species (population) extinction. He hypothesized that this situation was generally avoided because there were evolved mechanisms/adaptations at the species level (social behaviours) that guaranteed that the species density in the environment was always kept to an optimum. These mechanisms include certain types of social rewards, acquisition of territories, and so on *that act as cues to indicate whether the species density had grown over a threshold and required adjustment.* In Wynne-Edwards' view, epideictic mechanisms constitute real constraints on individual behaviour imposed by the species, and thus they were cases where (1) there had been some group selection at the level of the species favouring these traits; and (2) these traits constituted *engineering* or trans-temporally accumulated adaptations at the level of the species (populations).

Wynne-Edwards' ideas were soon attacked by Maynard Smith (1964) and Williams (1966). They believed Wynne-Edwards had confused genuine engineering, trans-temporally accumulated adaptations at the group or population level with behaviours that may be described as fortuitous benefits of engineering or trans-temporally accumulated adaptations at a different level.[4] In other words, Wynne-Edwards would have confused 'group benefit' with 'group engineering adaptation' (Lloyd 2021).

Generally speaking, engineering or trans-temporally accumulated adaptations at a given level of biological organization require a selection force detected by a selection model – a multilevel one, if the adaptation involves multiple levels – that will make it possible to produce such an outcome at the focal level. In such a case, several traits on the units at that level can evolve to serve a designed purpose for the form of organization that bears these traits – when it has evolved for having a specific function at the level – hence having the right type of accumulated causal history underlying their evolutionary origin (Brandon 1981, 1985; Sober 1984, p. 208, 1993, p. 85; Lloyd 2021). But a trait that benefits the group *now* is not necessarily an engineering adaptation at the

---

[4] Fortuitous benefits, under a more Evolutionary Change or Wrightian genetics approach, may arise in a variety of circumstances besides the engineering context, for example, if there is a non-zero selection coefficient for certain traits borne by the units at the higher level (see Wade 2016). To simplify our discussion in this section, we assume the Adaptationist School view according to which fortuitous benefits result from engineering adaptations at lower levels, or are product-of-selection adaptations.

group level. The trait may instead result from either a simple selection process at that level – constituting a product-of-selection adaptation – or be a lucky benefit at that level that results from the solution of a design problem at the lower level – thus having evolved with the right type of accumulated causal history or persistent response to selection at that lower level.

Williams perceived this last conceptual confusion as a fatal flaw in Wynne-Edwards' argument, which he described in terms of a lack of appreciation of the distinction between engineering adaptation and fortuitous benefit. Williams argued that the epidiectic mechanisms described by Wynne-Edwards, *if they existed in some species*, were merely fortuitous benefits for the species that had evolved as engineering or trans-temporally accumulated adaptations for the *individuals* of the species. Williams supported this claim by noting that within-species variation is very high for a persistent *response to selection* to accumulate at the species level, thus precluding the evolution of engineering or trans-temporally accumulated species-level adaptations. Organisms, on the other hand, are stable enough so that the variants they instantiate may persist for long enough to evolve engineering or trans-temporally accumulated adaptations at their level. In Williams' view, hence, the hypothesis that epidiectic traits had evolved via trans-temporal accumulation of adaptations at the level of the species was mistaken; they had evolved via trans-temporal accumulation of adaptations at the organismic level. The species does not constitute a unit of selection, Williams concluded at this time (later changing his tune in Williams 1992).

In the context of this dispute, Dawkins published his famous book, *The Selfish Gene* (1976/2006), where he introduced a conceptual framework that would soon be reformulated by Hull. Dawkins' framework is built upon the distinction between two different meanings of the expression 'unit of selection'. It can be used in the sense of the *replicator*, that is, an entity characterized by its longevity, fecundity, and copying fidelity; or in the sense of a *vehicle*, short-lived entities containing the replicators, and which the replicators use as an intermediary to foster their own success via the interactions of the vehicle with the environment (Dawkins 1982a, 1982b). *Replicator* and *vehicle* are abstracted away from the concepts of *gene* or *genotype* and *phenotype*. Dawkins believed that both types of units of selection were required to be present any time there is an evolution by selection process going on.

Dawkins' original reason for introducing this distinction – despite being also useful for disambiguating the controversy between Wynne-Edwards, Maynard Smith, and Williams, as the former would be talking about species as vehicles, while the latter two would be arguing that species are not replicators – derived from the way in which the concept 'unit of selection' was used in animal behavioural science and sociobiology. Many studies in these fields tended to assume what Dawkins called the 'Central Theorem of the Selfish Organism', according to which fitness maximization is attributed to the whole organism, rather than the genes carried by these organisms. Dawkins thought this to be mistaken and to have resulted from a tension inherent in the very concept of natural selection. This tension became especially salient after the discovery of the so-called junk DNA, series of base pairs of DNA that do not get translated into any protein; in fact, the existence of junk DNA was challenging in Dawkins' time, because it seemed to violate certain basic beliefs about the action of natural selection (see esp. Suárez 2019, p. 113).

The tension that Dawkins perceived can be described as follows. On the one hand, natural selection is *the source* of engineering or trans-temporally accumulated adaptations, that is, it is the mechanism that makes living entities look as if they had been designed *for/to fit* their environment. On the other hand, natural selection *requires* phenotypic variation to act, as phenotypic variance is the raw material on which natural selection can *select*.

Engineering adaptations, as Williams had shown, require entities with relatively little variation that could keep historical track of specific trait variants, that is, engineering, or trans-temporally accumulated adaptations, require replicators. Phenotypic variants, however, need to be in constant competition to be directly visible to natural selection, a point that Gould (1977, p. 24) illustrated in his review of *The Selfish Gene*. If replicators filled this latter role, then it seems that they would fail to fill the role of keeping historical track of specific trait variants.[5] Thus, the way in which replicators fill up the latter role is by making *vehicles*, short-lived entities that express different phenotypic variants and compete to foster the replication of their replicators. Replicators and vehicles constitute therefore two meanings of the expression 'unit of selection', and the two need to be carefully distinguished in debates about the units, because different researchers may be using the expression equivocally.

A few years later, Hull (1980, 1988a) reformulated Dawkins' framework by replacing the concept of the *vehicle* with the concept of the *interactor*. Hull introduces the two concepts as follows:

[I]n an effort *to reduce conceptual confusion*, I suggest the following definitions:

replicator – an entity that passes on its structure largely intact in successive replications

interactor – an entity that interacts as a cohesive whole with its environment in such a way that this interaction *causes* replication to be differential

With the aid of these two technical terms, selection can be characterized succinctly as follows:

---

[5] Note that this would not be always the case, as Hull (1988b) would illustrate with the cases of meiotic drive and segregation distortion (see later in the Element).

selection – a process in which the differential extinction and proliferation of interactors cause the differential perpetuation of the relevant replicators. (Hull 1988a, pp. 408–9; first emphasis added)

Hull's concepts provided more flexibility in capturing the idea that natural selection was a dual process that encompassed (1) the existence of fitness differences between different phenotypic variants at some level [interactor], and (2) the existence of hereditary or heritability processes that demarcated between cases of change due to selection and cases of change due to growth or phenotypic plasticity, and kept track of the underlying basis of the phenotypic variants, *eventually* fixing some of them [replicator]. *Interactor* and *replicator* always have different intensional meanings, but this obviously does not necessarily entail that they always stand for extensionally different objects. As a matter of fact, Hull's reason for preferring the concept of *interactor* over Dawkins' *vehicle* lay in the possibility of encompassing under interactor questions cases such as meiotic drive or segregation distortion where genes are simultaneously interactors *and* replicators (Hull 1988b, p. 28; Lloyd 1988/1994, pp. 124–5).[6]

Hull's introduction of the distinction between interactors and replicators can also be interpreted as a way of clarifying how the information contained in genetic models must be interpreted. Remember, this was the main reason why Dawkins introduced the bipartite distinction in the first place. In fact, most genetic models of interactors, either the philosophical or more technically framed, are grounded on the assumption that the expected fitness that will get enhanced is the overall fitness of the *interactor*, but its success is usually *measured* in terms of fitness parameters within the model that correspond to the fitness of *replicators* (or reproducers/reconstitutors).

This is important because it illustrates the *main rationale* for why the bipartite, interactor/replicator Framework was introduced: namely, to illustrate that even if the fitness parameters in some of the models representing interactors – especially those ignoring organismic trait values and their indirect genetic effects – stand for entities that make copies of themselves (replicators), the *causal* reason why their ability to make copies gets enhances or diminished is rooted in the phenotypic traits that are expressed in interactors that are in competition with each other and in the global environment (see Lloyd et al. 2008; Suárez 2019, pp. 119–25; Lloyd 2023) (Box 4).

---

[6] Note that, as it is expressed here, the concepts of interactor and replicator encompasses some properties that Lloyd (1988/1994) later isolates and attributes to the manifestor or the beneficiary of adaptation. In fact, one of the key problems with the split between only two units is that it is not enough to distinguish between the level where there is some selection coefficient and the level where an engineering or trans-temporally accumulated adaptation may evolve. It is important to bear this in mind to realize that the lack of these additional concepts kept triggering issues in the further debates about units.

Box 4 THE CONCEPTUAL NECESSITY OF THE ROLE OF THE REPLICATOR

While many of the recent developments in evolutionary biology demonstrate that the replicator, as it was originally defined, is not necessary for the evolutionary process, the same is not true about the *specific role* that replicators play in evolutionary biology, as well as in the bipartite and the tripartite frameworks.

Evolution by natural selection requires phenotypic changes and heritability, rather than just phenotypic change. Heritability can be measured and estimated using different tools, and its identification and representation is crucial to distinguish between cases of phenotypic change due to natural selection and cases of phenotypic change due to other factors, such as phenotypic plasticity. The introduction of the replicator is essential for demarcating cases where the phenotypic change is due to natural selection or to evolution, and cases where it is due to growth or developmental plasticity. Therefore, the concept of the replication, in its functional meaning later expanded to include reproduction and reconstitution, was an important contribution of Dawkins and Hull that needs to be preserved in contemporary analyses of units.

In conclusion, the specific properties associated to the replicator (copying fidelity, longevity, etc.) may not be necessary. This is well-documented in several cases of extended inheritance (e.g., epigenetic inheritance, symbiont heredity), which in turn enabled the introduction of the reproducer and, more recently, of the reconstitutor. Yet the functional role originally associated to the replicator still remains, being a key element in debates about natural selection, as well as a core meaning of 'unit of selection' that is sometimes conflated with the notions of interactor and manifestor of adaptation/type-1 agents that we isolate in this work.

Note that the bipartite formulation clearly distinguishes between an active element in the process (the interactor, which *causes* replication to be differential) and a relatively passive one (the replicator, which is copied).[7] Additionally, note that this bipartite framework can be used to redefine natural selection as the process in which the differential proliferation of interactors causes the differential replication of replicators, where the interactor and the replicator can but do not need to be one and the same object (Brandon 1988; Hull 1988b; Lloyd 1988/1994).

---

[7] The replicator (reproducer/reconstitutor) is not totally passive, because the action of replicating is itself an activity. Yet, the point that we want to emphasise here is that it is not actively in the process of selection itself—unless it simultaneously adopts the functional role of the interactor, in which case it would be active *in virtue of being an interactor*—rather in the process of copying.

A clarification is in order before ending this section. Since the mid-1990s and until today, the definition of the replicator has been questioned since it became clear that thinking of replicators as entities of which copies are made was insufficient to account for all selection processes. In many cases, the functional role is played by entities with two crucial characteristics: (i) they result from an event of transgenerational overlap (i.e., the entity in the previous generation transmits part of its body or material to the entity in the next generation), and (ii) they have the capacity of acquiring the capacity to reproduce through the process of development. For example, a sexual multicellular entity reproduces by the mix of two independent gametes which merge and develop into an adult sexual multicellular entity with the capacity of forming gametes that can be transmitted to the next generation. Following Griesemer (2000a), we will refer to these entities as *reproducers*. In other cases, though, transgenerational overlap is not even necessary, and the entity in the new generation results from the merge of many independent units, each of which retains its reproducing capacity. An example would be the formation of multispecies symbiotic consortia, such as the union of some animal hosts and their microbiomes. Following Veigl et al. (2022), we will call these *reconstitutors*.

Note that the expansion of the concept of the replicator into the concepts of the reproducer and the reconstitutor does not replace the necessity of the functional role that Dawkins and Hull originally attributed to replicators. It only changes the specific definitions of the concept that had been introduced in the previous literature. We will say more about the concepts of the reproducer and the reconstitutor in Section 4.

2.1.1 Relationship between the Recipe Approach and the Interactor/ Replicator Bipartite Framework

The bipartite analysis suggests an immediate bipartite interpretation of the recipe approach (Figure 1). The problem in the recipe approach that the bipartite framework amends was the failure to appreciate that the object that plays the causal role in the realization of heredity or heritability, and the object that expresses the different phenotypic traits that are systematically associated with the differences in fitness[8] are *not* always one and the same object – although they can be co-extensional sometimes, for example in the evolution of selfish genes (see Brandon 1988).

Another way of making the same point, suggested by Griesemer and Wade (2000), is by noting that Lewontin offers a structuralist or semi-formal approach, where what's relevant is the hierarchical nature of different selective

---

[8] The necessity of introducing the systematicity of the connection between phenotypic variants and their fitness has been analyzed by Millstein (2002).

Relationships between two schemas

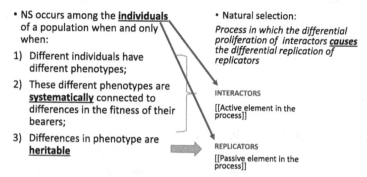

Figure 1 Relationship between the recipe approach (left), and the bipartite (interactor/replicator) framework (right).

Note that under the replicator/interactor framework, the entities specify functional roles that serve to guide the type of empirical evidence that needs to be found in nature, also including the mechanisms by which these roles are realized.

scenarios, whereas those endorsing the bipartite framework reinterpreted the analysis in functionalist or strictly formal terms. Griesemer (2005) also notes that the functionalist interpretation has the advantage that it is more abstract than the recipe approach insofar as it concentrates on the metaphysics of evolution, rather than on specific biological objects already known to the biologists, and in doing so:

> [it] analyzes the causal mechanisms through which genes and organisms play significant roles in the process of evolution by natural selection and subtracts kinds of explanatory factors which appear to rest on features peculiar to genes and organisms . . . The strategy is to abstract from the matter and structure of the concrete mechanisms of genes and organisms to yield a theory specified solely in terms of functions. The evolutionary process is regarded as having two components, genetic replication and selective interaction, and these are defined for any entities performing those functions. Since there may be distinct mechanisms for entities at genic, organismal, and other levels that satisfy the sort of causal structure required for evolution by natural selection to occur, the varying properties cannot be essential to the process. Such aspects are inessential to the roles of entities as relevant units in the evolutionary process. *Thus generalization (over the class of mechanisms and levels) is achieved by abstraction from the accidental properties of mechanisms at the gene and organism levels.* (pp. 72–73, emphasis added)

The functionalist interpretation allows therefore distinguishing between the units of interaction – and the hierarchy of units of interactions – and the units of heredity – a priori non-hierarchical if seen from a replicator point of view (but see Section 4). It also allows viewing these two abstract entities as specifying the *functional characteristics of the entities or mechanisms* that would be required to be found in nature to serve as evidence for the claim that an abstract unit is an interactor (unit of interaction) or a replicator (unit of heredity) (see Suárez 2019, pp. 119–25).

2.2 The Impact of the Interactor/Replicator Bipartite Framework

After its proposal, the bipartite analysis was in active use to apply to units of selection debates, which came to be considered debates about finding a hierarchy of interactors and the corresponding hierarchy of replicators. Additionally, it was used to frame several controversies, including the problems about group selection and species selection (e.g., Brandon 1988; Griesemer 2005) and, more recently, about multispecies systems (Dupré & O'Malley 2009).

Part of the work that came after Hull consisted in specifying criteria or methods to empirically identify interactors, as the criteria of 'cohesiveness' and 'causally biasing the replication of replicators', formulated as Hull did, were considered insufficient to establish whether units at a level fulfilled the role of interactors (Sober and Wilson 1993; Bourrat 2021), as well as ambiguous between two non-equivalent functional roles (Lloyd 1988/1994, 1992, 2023). The project of clarifying the meaning of the interactor was undertaken by several philosophers and biologists, who continued to add precision to the concept of interactor, making it one of the most precise concepts in the units of selection debates. Note that the project of spelling out clear criteria for identifying interactors becomes meaningful because it concerns distinguishing these organizational levels where natural selection is really acting from those where the perceived effects, even though *apparently* a result of selection, are merely cross-products of selection at other levels, or failures of the statistical tools used to measure selection (Lloyd 1986,1988/1994; Okasha 2006).[9]

For instance, Wimsatt (1980a, 1980b) defined the level of interaction in selection models in terms of heritable *context-independent* variance in fitness that causes the heritable variance in fitness at the lower levels to be *context-dependent*. Under this framework, if some lower-level units interact epistatically, then the adequate level to detect interactors is the higher level, where the results of epistasis, but not the epistasis itself, could be found.

---

[9] Note that the role of the replicator is not discussed in this section because most researchers trying to clarify the concept of interactor accepted the validity of the concept of replicator. However, the situation changed after the ETI project started. See Section 4.

Brandon (1982, 1985, 1988, 1990) tried to accommodate Wimsatt's conception with the generally accepted notion that selection acts on phenotypes. Brandon noted that epistatic effects occur among genes and are expressed at different levels, and combined this observation with the screening-off criterion formulated by Salmon (1971), concluding that an interactor can be identified at one level when the entities at that level show differential reproduction and '[t]he adaptedness [expected fitness] values of these entities screen off the adaptedness [expected fitness] values of entities at every other level from reproductive values at the given level' (1982, p. 319).

Lloyd (1988/1994) refined Wimsatt's criterion to introduce her 'additivity' criterion focusing on the characteristics of the variances in fitness or, informally put, that fitness is emergent at the higher level, but not at the lower level (Lloyd & Gould 1993; Gould & Lloyd 1999). According to it, there is a selective level of interaction at the higher level when there is additive variance in fitness at that level, and non-additive variance in fitness at the lower levels. In most cases, this would be due to epistatic interactions among the units at the lower levels (Wade 2016; Lloyd & Wade 2019).

Similarly, Sober (1984) and Sober and Wilson (1993, 1998) provided the criterion of *common fate*, according to which a group of entities is an interactor when the entities that compose the group are in fitness-affecting interactions mediated by a property that puts 'them "in the same boat"' (Sober & Wilson 1993, p. 551). The main idea was to distinguish cases of frequency-dependent selection, where Sober and Wilson believed interaction to occur exclusively at the lower level, from genuine group-level selection, which requires that the entities in the group are bound such that changes in the fitness of one affects the fitness of the other because the group *as a whole* competes against other groups. A similar idea has been recently put forward by Bourrat and Griffiths (2018), and also Bourrat (2019, 2021, 2023), who demand that the units composing the interactor relate to each other in terms of fitness alignment through fitness boundedness and contend that such realization is possible through epistasis between the elements.

Other approaches have centred the question on more technical aspects, concerning the type of statistical tools that are preferable in different selective scenarios to discover or support claims of interactors. For example, Sober and Wilson (1998) criticized Lloyd's additivity approach, based in the biological work of Arnold, Fristrup, Wade, and Lande, that identifies selection processes by tracking fitness correlations. They claimed that it failed to distinguish between cases of frequency-dependent selection and cases of genuine group-level selection. While Sober and Wilson (1998) would be correct if selection processes are looked at simply at one level (the level of the individual or lower level), where group and frequency-dependent selection are conflated with one another, they fail to appreciate that Lloyd's (1988/1994, 1994) additivity approach requires testing for selection *at two levels*. When doing so, the comparison of the group-level trait and its fitness would determine if it is a case of group selection (where one would perceive that variance in fitness is non-additive for interactors at the group level) or frequency-dependent selection (where this variance would be additive at the group level).

Another example of technical issues is present in the debate between Okasha (2006) and Goodnight (2015). Okasha (2006) agrees with Heisler and Damuth (1987) and Damuth and Heisler (1988) in saying that a multilevel selection approach is preferrable. He especially prefers to structure multilevel selection analysis by the Price approach, especially in the cases of soft selection, and by contextual analysis in other selective scenarios (cf. Goodnight et al. 1992). Goodnight (2015), also a multilevel selectionist, but one who comes from the evolutionary change school, disagrees about the broad applicability of the Price approach. Goodnight claims that contextual analysis, as introduced by Heisler and Damuth (1987) and Damuth and Heisler (1988) as an extension of the analysis of phenotypic selection originally advocated by Arnold and Wade (1984a, 1984b), should be generally preferable, even in cases of soft selection, a claim with which Lloyd agrees (2023).[10]

Overall, the issue of which methods and statistical tools should be preferred to detect interactors is still open and may depend on which research school the researcher comes from. But in general, it can be confidently argued that the identification of a level of interaction requires the combination of empirical evidence from fieldwork and laboratory experiments with technical statistical tools such as contextual analysis, or other techniques of regression analysis to detect whether there is a non-zero degree of interaction between the components. In other words, it requires gathering enough evidence that suggests that these processes are occurring at that level, even if they only do so for a relatively ephemeral time, and even if they do not lead to a process of engineering or trans-temporally accumulated adaptation or do not give rise to net evolutionary change as measured by the response to selection at the level under investigation (Okasha 2006, ch. 1; Lloyd 1988/1994, ch. 5).[11]

---

[10] This and the previous paragraphs are merely a summary of some approaches to show that the debate about the interactor was vivid, and one of these debates where the concept became more precise over time, despite some researchers recently attacking this impression of the concept (Section 4). For a more expansive review, see Okasha (2006) and Lloyd (2023).

[11] These philosophical disputes about what constitutes 'enough evidence' and how to test different claims about detecting a unit of selection at a certain level had an important influence on working biologists. For example, Williams (1992) pursues a hierarchical model of selection after having been convinced by Lloyd (1988/1994) that the detection of engineering or trans-temporally

2.3 Clarifying and Expanding the Interactor/Replicator Bipartite Framework into a Tripartite Framework

Given that our purpose is not to review different conceptions of the interactor to determine which one is better, but rather to articulate and defend the usefulness of the DP for the study of the debates about units, we now introduce an amendment to the bipartite analysis which isolates the meaning of a third functional role: namely, the manifestor of (engineering) adaptation (Lloyd 1988/1994, 1992, 2023). This third meaning of the expression 'units of selection' expands the bipartite framework into a tripartite one and is the framework we advocate in this Element. This role was originally introduced by Lloyd in the context of what we have called the DP about the units,[12] but similar concepts have been later re-isolated by other authors in the context of different biological disputes about the units, possibly the most salient being Okasha (2018), who introduced the similar concept of 'type-1 agent'.

A key point to note about this specific meaning that the expression 'units of selection' adopts is that it is the usual meaning which, in combination with the interactor, is presupposed in the adaptationist achool to the study of natural selection (see Section 3). This is the main reason why both Lloyd and Okasha argue that the concepts they introduce frame the adaptationist side of the controversy – even though the concept of the manifestor is also of use for those working under the lens of the evolutionary change school.[13]

After its proposal, the tripartite analysis was in active use to apply to units of selection debates. For example, it was used to frame several controversies, including the problems about group selection (Lloyd 1988/1994, 1994; Williams 1990; Grantham 1994; Maynard Smith 2001), to help solve problems in species selection (Lloyd & Gould 1993; Gould & Lloyd 1999; Jablonski and Hunt 2006; Jablonski 2008), as well as multispecies selection (Lloyd & Wade 2019; Suárez & Triviño 2020).

Under the tripartite approach, the manifestor of adaptation is an entity in the biological hierarchy that bears traits that make it look as if it were 'engineered'

---

accumulated adaptations at one level is independent from the detection of interactors at that level. Similarly, Biernaskie and Foster (2016) provide evidence suggesting that group selection may play a role in explaining aggression in spider colonies but are careful enough not to claim that multilevel selection or kin selection explain the observed behaviours better, after being convinced by Okasha (2016) that the two approaches are not causally equivalent.

[12] Lloyd's (1988/1994; 2023) original analysis also included a fourth meaning, the beneficiary, that we do not consider in this work, as one of us is not sure about its relevance in today's research on units.

[13] Okasha does not subscribe to the DP, and in fact he does not really discuss the relationship between his concept of type-1 agency and the units of selection controversy, arguing that it would be hard to understand the relationship between both projects. We think the approach we offer in this Element covers that gap.

or tinkered to fit or respond to problems/challenges in its environment. These traits can be called *engineering* or *trans-temporally accumulated* adaptations, as they are traits that result from the cumulative transgenerational effect of natural selection acting at a specific level, showing a high degree of 'cohesiveness' or organization (Lloyd 2021). Note that the type of cohesiveness of manifestors is due to the existence of at least one (possibly more) engineering trait at the level. This contrasts with the type of 'cohesiveness' shown by ordinary interactors, which is customarily reduced merely to the existence of some indirect genetic effects between the components of the interactor not tied to the existence of emergent adaptations (see Section 4.1 for the problem of the 'cohesiveness' of the interactor).

Note that this meaning of 'manifestor' would have been confounded with the meaning of 'interactor' by Hull in his original bipartite framework. He explicitly states that he considered the question of finding higher-level interactors to concern 'whether entities more inclusive than organisms exhibit [trans-temporally accumulated] adaptations' (Hull 1980, p. 325). This requirement would be characteristic of the manifestor of adaptation, rather than just the interactor. However, Hull simultaneously characterizes interactors as the entities that *cause* replication to be differential (Hull 1988a, p. 403), a requirement that is characteristic of what we have referred to as *interactors*.

Sober and Wilson (1993) present another good example of this conflation between interactor and manifestor of (engineering) adaptation, when they require interactors to exhibit a level of *functional organization* – usually a requirement for adaptation – enough to be considered units of 'common fate'. This conflation was rejected by Grantham, as well as Smith, both of whom wrote commentaries responding to Sober and Wilson's paper. Grantham (1994, p. 623), for instance, claims that 'if we recognize that group selection does not require functional organization (either as a necessary prerequisite or a necessary consequence), then the failure to find such organization is not definitive evidence against group selection'. Smith (1994, p. 636) is even more explicit when he states that Sober and Wilson are 'conflating selection acting on groups [as interactors] with selection of group adaptations [manifestors]'.

Being a manifestor of adaptation at a given level, hence, requires the possession of emergent traits in the form of engineering or trans-temporally accumulated adaptations at the specific level, a requirement which is exclusively of the manifestor and not necessary for interactors, where only the fitnesses need to be emergent (Lloyd 1988/1994). To put it another way, detection of an interactor at one level requires the detection of a product-of-selection adaptation at that level – a trait-fitness causal covariance at that level – as well as other statistical relations at other levels. But it does not necessarily require or need the detection of a designed, engineered or trans-temporally accumulated adaptation at that level – a requirement, instead, for finding a manifestor at that level (Lloyd 1988/1994).

The concept of manifestor of adaptation has recently been reintroduced by Okasha (2018) under the expression 'type-1 agency' in the context of his work on agential thinking in evolution. Okasha (2018, p. 5) defines his concept as follows: 'type 1 [agency] is a legitimate expression of adaptationism, but it relies on a crucial presupposition. It presupposes that the entity that is treated as an agent exhibits a "unity-of-purpose," in the sense that its evolved traits contribute to a single overall goal'. And he specifies '[f]or an entity to be treated as an agent, for the purposes of adaptationist theorizing, its evolved traits must contribute to a single overall goal, and thus have complementary rather than antagonistic functions' (2018, p. 52). Defined this way, type-1 agency closely aligns with the optimization methods that are frequently used by kin selection theorists in their efforts to understand what type of forms will or could be produced by natural selection, just as Goodnight describes the Adaptationist School framework (Goodnight 2015), a point we will expand on in Section 3.

Note that Okasha's definition is stricter than the definition of the manifestor we have introduced before, as a type-1 agent must have a unity of purpose and possess a set of engineering or trans-temporally accumulated adaptations. For this reason, it is not an accident that Okasha exemplifies type-1 agency by talking about cases where there is an almost total suppression of within-organism or lower-level conflict, a requirement which is characteristic of the adaptationist school but not necessary for the manifestor of adaptation. So, properly speaking, and even though we consider them as almost synonymous given both aim to capture the type of unit used by adaptationists, the type-1 agent must be considered a subclass of the manifestor.

Note, however, that Okasha is cautious enough to show that optimization models are not always possible, feasible, or even desirable, because there may be multilevel selection scenarios where evolution by natural selection occurs but does not lead to the formation of type-1 agents. To quote:

> This is not to say that group or multi-level selection is rare, but only that *it does not usually lead groups to exhibit the degree of internal harmony that a typical individual* [manifestor/type-1 agent] has. Indeed, in a sense this is a definitional rather than a substantive truth, since where groups do evolve a high degree of cooperation and functional integration, we tend to elevate them to the status of "individuals" and regard their members as parts of a single whole. (Okasha 2018, p. 53, emphasis added)

Note that this paragraph speaks in favour of making a distinction between the interactor and the manifestor/type-1 agent, a distinction that Lloyd (1988/1994, 1992) had previously advocated, and that we endorse and defend in this work. Questions about interactors would not immediately concern the study of optimization of traits, or the appearance of engineering or trans-temporally maintained adaptations at one level, but rather the evolution of any trait, as a mere product-of-selection, at that level (Wade 1979, 2016; Goodnight & Stevens 1997; Goodnight 2015). Okasha (2006) also characterizes these models that concern questions about interactors as distinct from the maximizing or type-1 agency models concerning questions about the manifestor, and we think that keeping the two meanings apart is useful for today's biology. In discussing trait-group models such as D. S. Wilson's (1975), Okasha (2006) contends that these models are cases of multilevel selection to study how a trait changes among the components of a group in virtue of their interactions within the group. Okasha admits that these are genuine selection models, that is, they track genuine questions about the units of selection, even though they do not track questions about type-1 agents. We agree, but we clarify, that they are genuine models *to answer interactor, not manifestor, questions.*

Turning to empirical examples of manifestors, Lloyd (1992, 2021) illustrates the concept with the case of the beaks in finches, as studied by Grant and Grant (1989, 2020). It contrasts with the evolution of industrial melanism in *B. betularia*, which would be, on the other hand, a mere product-of-selection adaptation – a case of selection for a trait (interactor) *without a history of design or tinkering.* Suárez and Triviño (2020) illustrate the manifestor by relying on the evolution of the bearded sting in honeybees, a trait that constitutes what Haldane (1932/1990) would have called an *altruistic* trait. The bearded sting in honeybees is clearly a result of cumulative selection, hence it is an engineering or trans-temporally accumulated adaptation at some level. Yet, it would be perplexing to think of it as an engineering adaptation at the individual organismic level, given that it is a structure whose biological function – its use to protect the colony from attacks – causes the death of its bearer.[14] The perplexity disappears once one realizes that the adaptive function of the bearded sting in honeybees is to protect the colony *at the expense of the lives of the individual bees bearing the trait.* The colony, that is, the higher level, is therefore the manifestor of the

---

[14] Note that the biological function of a trait is, in the context we are discussing here, a combination of its etiological and dispositional functions, that is, if a trait has a biological function, it does what it has evolved to do (Lloyd 2021; Suárez & Triviño 2020, pp. 2–4, and references therein).

adaptation in this specific case, whereas the trait constitutes a fortuitous 'damage' for the individuals bees bearing it.

2.3.1 Correspondence between the Tripartite Framework and the Recipe Approach

The tripartite distinction can also be put in correspondence with the recipe approach (Figure 2). In the tripartite approach we have presented, the interactor corresponds to the object whose phenotypic differences are systematically – causally – connected to its fitness (emergent fitness; see Goodnight 2015; Wade 2016). So, in a way, it corresponds to a combination of the Recipe's criteria (1) and (2). This model can fit quite generally, if we take the 'individual' interactor as a higher level of organization than an isolated cellular or multicellular organism. For example, take a multilevel selection process on which interactors can be at any level; an empirical way of detecting the presence of an interactor may depend on detecting certain biases in the expected heritability of replicators, which result from the epistatic or indirect genetic effects at the lower level which allow for the action of selection at a higher level (see Goodnight 2015; Bijma & Wade 2008; Bijma 2014; Wade 2016).

Hence, in the tripartite approach, the interactor *causes* the differential replication of replicators in virtue of its trait-mediated differential proliferation. Thus, *the real interactor question in evolutionary biology is identifying which traits at which level of entity are interacting with their environments such that replication is differential* – or reproduction, in reproducer models, and reconstitution, in reconstitution models (see Section 4). Simply, at what

Three types of units of selection

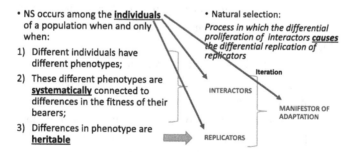

Figure 2 Relationship between the recipe approach (left), and the tripartite framework (right).

Note that in future approaches, the tripartite framework would also include the Reproducer and the Reconstitutor.

level of entity is all this *selection* action occurring with its environment?[15] This entity is a unit of selection functioning as an *interactor* in this particular evolution by selection process under study.

Additionally, the tripartite schema requires a unit that plays the functional role of the replicator, that is, an entity of which copies are made (but see Section 4 for our endorsement of the concepts of the reproducer and the reconstitutor). Replicators are simultaneously necessary to speak of selection processes, and they need not be at the same focal level. The necessity of both functional roles is directly inferred from what is represented in the genetic and evolutionary models, whether formal or informal, as well as of the necessity of distinguishing between phenotypic changes due to growth or plasticity and those due to evolution by natural selection.

Finally, the schema includes the manifestor/type-1 agent which can be a very specific product of these processes, in cases where the iteration of the processes takes a very specific form. Finding a manifestor/type-1 agent is then not necessary for inferring the existence of a selection process, *a very important fact to keep in mind for all concerned.* Finding a manifestor is, however, *sufficient,* for inferring the current or past presence of an interactor in the selection process, as one may have existed in the very process that produced the engineering adaptation. This allows asserting that units at the manifestor level have been tinkered/engineered, through selection of the interactor, and might likely continue to be so, assuming certain conditions about the inheritance of environments and traits.

3 Two Sources of Misunderstanding in Past and Today's Debates about Units

As the previous discussion clearly illustrates, our main claim in this Element is that the DP is still necessary. This is sometimes not recognized in the philosophical and biological literatures because of two types of misunderstanding in debates about units. In this section, we introduce these two main sources of misunderstanding and confusion.

On the one hand, the expression 'unit of selection' has been and still is used to mean at least the three different types of non-co-extensional functional concepts that we isolated in Section 2: namely, interactors, replicators/reproducers/reconstitutors, and manifestors of adaptation/type-1 agents. These three concepts are distinguished by the *different types of functional roles* that biological objects in different organizational levels can occupy, and the *different research questions*

---

[15] Note that multiple entities at multiple levels can be identified for any selection scenario under investigation. Classically, the t-allele case in the house mouse exhibits three levels of interactor simultaneously (Lewontin 1962).

that these distinct roles trigger. Even though one object can simultaneously occupy the three roles, and some contemporary biological research – most noticeably, research on the ETI (see Section 4) – asks combinations of questions about how a specific object in the biological hierarchy satisfies the three roles at the same time, the three roles must be theoretically and conceptually distinguished, even though questions combining two or more of these roles are occasionally asked in combination. This is because many of today's debates in biology still require the distinctions to be explicit to prevent biologists from talking past each other – specifically, when they are asking these different questions *independently.*

On the other hand, another foundational source of biologists talking past one another lies in their using very different frameworks of multilevel selection theory to shape their research, and its goals and purposes. Goodnight and Stevens (1997), and later Goodnight (2015) and Wade (2016), have, we think correctly, described these frameworks as distinct 'Schools' in Evolutionary Thought. One school has its source in Wright's work from the 1930s, which they call the *experimentalist, genetic, Wrightian,* or *Evolutionary Change School* in multilevel selection. The other, frequently more informal one, arises out of Haldane's ideas and challenges about the evolution of 'altruistic' traits, from the 1940s, which they call the *Adaptationist School* – sometimes also called *Fisherian School*. To quote:

> The genetic [Evolutionary Change] school is experimentally oriented, with experiments derived from the field of quantitative genetics. It focuses on *processes,* such as *changes in the distribution of phenotypes over generations,* rather than the *level* of biological organization at which *traits are adaptive.* (Goodnight & Stevens 1997, p. S61, emphasis added)

Such evolutionary change theorists may use multilevel selection (MLS) theory to study, for example, how certain phenotype-fitness connections at a certain organizational level that affects the pattern of heritability at the same or at a lower level also affects the distribution of traits. In this sense, it was devised to study *the process of selection itself* (Goodnight & Stevens 1997).

A typical evolutionary change research problem would encompass finding a selection process acting at a specific level by finding a bias in the hereditary pattern and trying to determine how the trait would change over time due to the existence of a selection process acting at that level. Researchers working in this school are usually concerned with interactor questions, although may sometimes also study engineering or trans-temporally accumulated adaptations at any level. As Lloyd (1986, p. 386) once described a typical goal of the evolutionary change school: 'Questions about interactors [as units of selection] focus on the description of the process itself – the interaction of entity and environment, and how this affects evolution – rather than on the outcome [engineering or trans-temporally accumulated adaptations] of this process.'

In contrast to those guided by the evolutionary change school, a researcher working in the adaptationist school 'looks at *patterns* and attempts to infer *processes*' (Goodnight & Stevens 1997, p. S60, emphasis added).

In some cases of most interest to the adaptationist school, they use kin selection models – a version of hierarchical selection models – to attempt to establish *how* engineering or trans-temporally accumulated adaptations *could possibly* evolve against the evolutionary interest of the entities at the lower level of biological organization. These higher-level engineering or trans-temporally accumulated adaptations are usually traits that 'benefit' the higher-level or 'altruistic' traits (Okasha 2003). Many of their projects revolve around solving the puzzles of how the emergence of these high-level manifestors of adaptation could possibly happen. Using optimality modelling and seeking stable equilibria in kin selection, inclusive fitness, or game-theoretic models to discover the higher-level manifestors/type-1 agents, they search for genetic models that could explain the evolutionary emergence of such 'altruistic' traits (see Goodnight 2015; Goodnight & Stevens 1997; Maynard Smith 1978; see also Wade 2016).

A typical adaptationist research problem would encompass two research questions:

1. Identification of a higher-level trait that is optimal at that level but seems to go against the selective interest of the entities at the lower level (genes, organisms, independent lineages), making the trait unlikely to evolve;
2. Imagination of a selective scenario where this trait could have been evolved against the odds.

Given that philosophers (and some biologists) have worked for decades developing definitions of 'Adaptationism' (Orzack & Forber 2017), it may seem strange to refer to a single adaptationist school here, in the context of hierarchal or multilevel genetics theory. So we will pause for a few moments to consider a couple of relationships between this specific form of adaptationism we refer to here, and some of the other primary meanings that have been discussed in the wider literature. We would like to emphasize the distinction between empirical, explanatory, and methodological/heuristic adaptationism, and how these different versions map onto the adaptationist/evolutionary change school approaches we have distinguished.

A first approach to adaptationism is 'Empirical' adaptationism, the view that natural selection alone gives a *sufficient* explanation for most traits, and that any other potential evolutionary factors play little role in evolution of such traits locally (see Orzack & Sober 1994; Charnov 1982; Reeve & Sherman 1993; Amundson 2001; Godfrey Smith 2001; Sansom 2003; Lewens 2009). This approach – often arising in animal behaviour – does recognize that there may be a variety of other evolutionary *possible* factors (Box 1). However, the claim is that they play little role in *actual* evolutionary outcomes (see discussion in Pigliucci & Kaplan 2000; Amundson 2001).

Empirical adaptationism is frequently combined with permissive views of the concept of *adaptation*, as opposed to the most traditional views that defined a trait as an adaptation in relation to its engineering or trans-temporally accumulated history (see Burian 1992; Lloyd & Gould 2017). Reeve and Sherman, for example, redefine adaptation: 'An adaptation is a phenotypic variant that results in the highest fitness among a specified set of variants in a given environment' (1993, p. 1). On Reeve and Sherman's 'current fitness' definition of adaptation, the research question of an empirical adaptationist asks only about the current fitness, not about any trait's history of selection, a move that sweeps a much larger group of traits into the category of adaptation (see discussion of the relationship between definition and quantity of resulting adaptations in Lloyd (2021, pp. 82–4)).

A second approach is 'Explanatory' adaptationism, which claims to define the ultimate goal of evolutionary science: namely, evolutionary biology's central goal is to give adaptationist explanations for *all* the traits in an organism. In Gardner and Welch's words: 'The cardinal problem of biology is to explain the process and purpose of adaptation' (2011, p. 1801). In other words, evolutionary biology should concern itself primarily with giving explanations of the existence of traits in virtue of the engineering function they evolved to play. Explanatory adaptationism assumes that explanations of the origins of traits in terms of their adaptive function are the best kind of explanations (see Maynard Smith 1978; Charnov 1982; Mayr 1983; Reeve & Sherman 1993). The notion of adaptation used in explanatory adaptationism is highly compatible with the notion used in empirical adaptationism, and it is way less restrictive than the engineering/trans-temporally accumulated notion used in methodological adaptationism, which is the position to which we turn next.

'Methodological' or 'Heuristic' adaptationism assumes that a trait under study has been a target of a selection process. Be it a gene, cell, organism, family, group, or lineage, one starts by assuming that the trait is serving a particular adaptive function for the target of selection, which we call an evolutionary engineering or trans-temporally accumulated adaptation. This is the most prominent meaning of adaptationism in the evolutionary theoretical literature, as well as in biological practice (Maynard Smith 1978; Bock 1980; Mayr 1983). It is also the form of adaptationism that underlies the adaptationist genetic research programme of multilevel selection – adaptationist school in evolution – that we utilize in this Element (Lloyd 2005, 2015, 2021; Green 2014; Suárez & Triviño 2020).

As Lloyd (2015, 2021) has shown with her 'logic of research questions' approach, the type of questions posed by methodological Adaptationists – 'what is the function of this trait?' – commits them to, de facto, finding an engineering or trans-temporally accumulated adaptation story for the existence of every trait. This entails that a methodological adaptationist would proceed with research into a specific trait by exploring the space of possible selection processes that might have produced the *allegedly* functional adaptive trait in question, resulting in an explanation that claims 'this trait is an adaptation for such and such an evolutionary function, resulting from a history of accumulating selection for this function over evolutionary time' (see discussion in Lewontin 1978; Mayr 1983; Williams 1966; Lloyd 2021). Suppose such alleged function is P, and that a methodological adaptationist fails to find support for the first set of adaptive possibilities supporting P. Then she is obliged to come up with other possible selective scenarios, resulting in a new outcome, for instance, suggesting that the function of the trait is Q, R, S, and so on. Lloyd (2015, 2021; see also Gould and Lewontin 1979) contends that this procedure reveals a primary weakness of methodological adaptationism; alternative adaptative answers may be created ad infinitum, *before* alternative evolutionary explanatory causal factors are, in actual practice, considered seriously or added in combination. This also shows that methodological adaptationists are usually committed to a form of empirical and explanatory adaptationism, as well.[16]

The main take-home message is hence that, even though adaptationism can take different forms, it can be characterized as a general school in evolution centring its research questions about units on finding what we have called engineering adaptations and their manifestors/type-1 agents. As posited by the DP that we advocate here, the adaptationist interest in finding manifestors/type-1 agents contrasts with the type of units investigated under the evolutionary school of evolution. The former are searching for interactors that are also manifestors, while the latter are most often studying plain interactors. This gives rise to unfortunate results, as Goodnight has summarized:

---

[16] Some evolutionary biologists operate accepting all three of the above forms of Adaptationism (Dawkins 1976; Maynard Smith 1978), while other biologists reject all of them (Gould & Lewontin 1979; Wagner et al. 2000; Carroll 2005), and the rest are in between (see discussion in Orzack & Forber 2017).

Unfortunately, these differences in approach and language have tended to isolate the kin selection [Adaptationist] and MLS [Evolutionary Change] traditions. . . .

This failure of communication between these two approaches has had the consequence that important results from the MLS literature do not inform research following the kin selection tradition . . . The result is that researchers are unaware of how important the results of the other fields are to their own research. (2015, p. 1743)

With these distinctions under our belt, we have two key messages: (1) neglecting the polysemy of the expression 'units of selection' is not a profitable research avenue, because it conflates a series of theoretically and conceptually independent research questions with each other, as well as conflating the type of evidence that must be gathered to reply satisfactorily to each of these research questions by ignoring fundamental nuanced differences in biological practice; (2) the criticisms of the DP are based on several misconceptions about the rationale of the project, which fail to appreciate its real nature, goals, and effectiveness (or 'success'), and may ignore significant differences between schools of evolution.

3.1 Today's Controversies Where the Tripartite Framework Is Required

3.1.1 Interactor versus Replicator: The Case of Indirect Genetic Effects

The necessity of distinguishing the interactor from the replicator roles in genetic models, and knowing the real source of the causal information that affects the success of replicators (i.e., their fitness values), has been recently noted by geneticist Bijma (2014). He denounces the tendency to ignore interactor questions concerning indirect genetic effects (IGEs) in the kin selection literature from the adaptationist school we discussed (see Section 3).

Again, IGEs include the interactions between phenotypic traits specified by those genes in different organisms or groups that influence genetic success values (i.e., genic or genotypic fitness). They can appear when organisms' phenotypic traits interact with one another in a cooperative (or competitive) way and may thereby increase one another's reproductive success through these phenotypic trait interactions, or genetic fitness values. Bijma complains that kin selection modellers (adaptationist school researchers) tend to ignore both these phenotypic trait interactions and their genic fitness consequences, despite their well-documented genetic impacts.

In the group selection vs kin selection debate, for example, effects of genes are usually specified entirely *in terms of fitness cost and benefit of the interactions* [replicator], whereas *the effects on individual trait values are disregarded* [interactor] (for example, Gardner et al. 2011). A fundamental principle in genetics, however, is that the genotype affects the phenotype and the phenotype subsequently affects fitness (Lande & Arnold 1983). The ongoing fitness-centred [kin selection-centred] debate hinders scientific progress, both in theoretical and empirical studies ([empirical study:] Reeve and Keller 1999; [theoretical study:] Okasha 2010).

Because fitness-based [kin-selection] models disregard IGEs on trait values, such models do not predict response in traits subject to IGEs (for example, Gardner et al. 2011). This is a severe limitation, as the effect of competition or cooperation on fitness of individuals will usually work by effects on their [phenotypic, i.e., interactor] trait values, meaning that not only fitness but also traits are subject to IGEs. (Bijma 2014, p. 69, emphasis added)[17]

In Bijma's formulation, both the replicator and the interactor are functional concepts – as they were in Hull and are in this Element – and therefore biological research can be understood as discovering specific biological entities that fulfil each of the roles. Hence, *discovering new levels of selection consists in finding out the empirical evidence and the specific mechanisms that fulfil, independently or in combination, each of these roles.* Conflating these roles with one another would be a serious mistake.

3.1.2 Interactor versus Manifestor/Type-1 Agent

A first scenario where interactors must be distinguished from manifestors/type-1 agents concerns the construction of genetic models. In some cases of most interest to the adaptationist school, especially in kin selection models, engineering or trans-temporally accumulated adaptations have evolved against the evolutionary interest of the objects at the lower level, as traits that 'benefit' the higher-level, or 'altruistic' traits, to express the idea by using the terms previously used by Williams and Maynard Smith. Many of their projects revolve around solving the puzzles of how the emergence of these manifestors of adaptation or type-1 agents could possibly happen. Using optimality or kin selection modelling and seeking stable equilibria, they search for genetic models that could explain the evolutionary emergence of such 'altruistic' traits against the odds (Goodnight & Stevens 1997; Sober & Wilson 1998; Lloyd 1999; Goodnight 2015). But evolutionary change researchers, which largely rely on multilevel selection – and not kin selection – models, have persistently shown that the presence of an interactor at the higher level may be feasible even if there is a high degree of competition and variation between the entities at the lower level. In fact, a high level of competition at the lower level does not

---

[17] Note that Bijma writes assuming an Evolutionary Change approach. However, the distinctions being made in this passage are valid for both approaches.

necessarily preclude the efficacy of selection at the higher level. Evolutionary change researchers have insisted that their results on this topic are relevant for those working under an adaptationist framework: ' this [experimental] study illustrates that group selection is more favorable for the evolution of social behaviors than is kin selection. To the extent that morphology, physiology, and development affect the manifestation of social behaviors, then these traits will also be influenced in their evolution by population structure' (Wade 1980b, p. 854).

This is an example showing that group and kin selection models normally rely on different model assumptions about population structure and intergroup selection (Hamilton 1975; Wade 1980a, 1985; Lloyd 1988/1994). These diverging assumptions are related to the fact that two different kinds of units – interactors versus manifestors/type-1 agents – are being modelled. These assumptions also translate into the dynamics of the models, since the kin selection models used by the adaptationist school to discover the existence of a manifestor at a level do not necessarily produce the same number or character of equilibria as the evolutionary change multilevel selection models (Lloyd & Feldman 2002; Lloyd et al. 2008), nor are they causally equivalent (Okasha 2016), despite the repeated claim that kin and multilevel selection models are mathematically equivalent (Dugatkin & Reeve 1994).

Based on this, we contend that this specific meaning of the expression 'units of selection' to refer to manifestors/type-1 agents is mostly in use by those working under the adaptationist school. Note that the very intuitive notion of this trans-temporally accumulated engineering adaptation manifested as a unit of selection does not appear in the classical recipe approach. Users of the recipe approach can see that two or three of their requirements can be met by this single functional role of the manifestor of adaptation, and are often conflated with it. This leads to a conflation of the manifestor/type-1 agent with the interactor. Finding interactors and tracking their evolutionary process constitutes one of the main interest of those working on the evolutionary change school in evolution. Given their preferred approach, these researchers rarely model manifestors/type-1 agents. As interactors do not need to fulfil the same type of requirements as manifestors/type-1agents, the existence of these two types of units supplies important support for the preference of the DP in its tripartite version over the recipe approach or the bipartite version of the DP.

It is important to realize though that members of the adaptationist school frequently ask questions about interactors and manifestors/type-1 agents *in combination, aspiring to model and empirically find entities that fulfil both roles simultaneously.* To put it differently, they look for entities that are experiencing a process of selection now, and whose evolution is a product of past selection producing engineering adaptations at that level(s) (see Lloyd 2023 for sources). Ontologically and epistemologically speaking, however, the fact that a research tradition asks these two questions in combination, looking for objects that simultaneously satisfy both roles, does not entail that they are not distinguishable questions, discriminating between different types of roles that different objects may fulfil differently.

In addition to this, the lack of clear meanings for the expression 'units of selection' leads to disputes resulting from failures in communication between the members of these different schools of evolution. When these disputes arise, it is imperative for good science that the relevant research questions are distinguished, along with the type of evidence required to answer these research questions. We further endorse the necessity of such a distinction in this work, and claim that taking the two concepts into account is important *even when one is asking both types of questions simultaneously.* This is one reason that the DP was introduced, and a task that the framework is ideally suited for.[18]

Recent biological work illustrates how the concept of *manifestor* is frequently conceptually *conflated* with and *collapsed into* the (engineering or transgenerationally accumulated) adaptive-neutral concept of the *interactor*, and how biologists themselves are conscious that this concept must be distinguished for better communication with one another. One example of this has been codified by Gardner and Welch (2011), who have offered a 'Formal theory for the selfish gene'. They formalize the relationship between gene-centred population genetics models, using the price equation, and an agency-based approach to adaptation (related to Okasha's later agent-based approach (see Section 2) based on optimization of phenotypes produced by alleles. In other words, they are offering formal models of both parts (1) and (2) of the adaptationist project mentioned in Section 3, with the *stipulation* that it is all reduced to a causation based on the selfish gene.

In one highly significant feature of their theory, Gardner and Welch claim that their models provide justification for excluding non-additive genotypic interaction effects from their evolutionary models. But these non-additive (otherwise known as epistatic) interaction effects are *the very interactions* that drive both *higher-level interactor selection,* and any *higher-level adaptation* that might occur in the evolutionary change multilevel selection models. Gardner and Welch justify the exclusion of such dynamically important features on the

---

[18] Evidence of how the DP, in its Tripartite formulation, played a role in solving debates among biologists can be found in Lloyd (1988/1994; 2001). Good examples of controversies that were partially solved by adopting the DP include the cases of species selection (Jablonski 2008; Gould & Lloyd 1999; Gould 2002; Williams 1992) and group selection (Williams 1990; Maynard Smith 2001; Wade 2016; Lloyd & Wade 2019).

basis that such effects 'show no correspondence with the optimization program' (2011, p. 1809). Thus, while they acknowledge that 'neglect of these epistatic effects will lead to a less complete (and potentially quite inaccurate) account of the evolutionary process ... the gene's eye view *does* function adequately as a theory of adaptation' (2011, p. 1809; emphasis added).

Note that this thus implies that, on their formal theory for the selfish gene, no adaptations of interactors could even theoretically appear at the higher level of interactors. Interestingly, Gardner and Welch utilize the DP, and emphasize that *they are not asking research questions about interactors*, that is, they state that they are not asking about 'the manner in which biological entities interact with their environment in a way that makes a difference to replicator success (Hull 2001; Lloyd 2001, 2005; Okasha 2006)', but rather, questions about the man-ifestor/type-1 agent, or what they call 'ultimate causation': 'why do we observe adaptations, at what level of biological organization are they manifest, and what do they appear designed to achieve (Gardner 2009)?' (2011, p. 1810).

They conclude that their notion of 'maximizing agent', [manifestor/type-1 agent], focuses on function, intention, or design, *rather* than 'target of selec-tion', or interactor, thus zeroing in on the significance of this distinction (Gardner & Welch 2011; Lloyd 1988/1994). Most importantly, this also shows that methodological adaptationists are committed to a different concept of 'unit of selection' than the one used by those working on the evolutionary change school of evolution. As a result, the former are committed to a different view of natural selection and evolution in general, that is one where the evolutionary process is completely identified with Darwin's process *tout court* (Box 1), rather than with the entire set of evolutionary factors.

3.2 Summary of the Tripartite Version of the Disambiguating Project and How It Disambiguates

All the evidence we have shown clearly illustrates the relevance of the DP to clarify or solve some of *today's* controversies in the units of selection debates. Failing to note that some of the controversies appear because different authors mean different concepts when they use the expression 'unit of selection' creates serious but unnecessary disputes. The tripartite framework, which we advocate in this work, disambiguates three of these meanings. "Units of selection" can be used to refer to interactors, replicators/reproducers, or manifestor/type-1 agents, and sometimes biologists even mean combinations of these three concepts (e.g., sometimes manifestor and interactor, or manifestor and reproducer questions are addressed simultaneously; Lloyd 1988/1994). Table 1 summarizes the main types of questions that we have disambiguated under the tripartite version of the DP.

Table 1 What is a unit of selection and how to identify it?

Type of unit	Research question it captures	Type of evidence	Modelling practices used to study these questions	Predominant context where questions about this type of unit are asked
Interactor	What entity is interacting as a whole with its environment such that replication/reproduction is differential?	Differential replication or reproduction *at the focal level or at the lower – and higher – levels* measured by statistical analysis of empirical evidence (fieldwork, laboratory). The purpose is to discover IGEs between the focal-level units genuinely biasing the reproductive pattern	Multilevel genetics models, both in inclusive fitness models and full population genetics models, multilevel selection 1/multilevel selection 2, contextual analysis/price equation	Experimentalist tradition in evolutionary thinking/genetics; quantitative genetics; evolutionary change approach; adaptationist optimality modelling; paleobiology
Manifestor of adaptation/type-1 agent	For manifestor and type-1 agent: Can we detect a history of selection on a trait or traits of this entity that contributed to its emergence over time as an accumulated, seemingly engineered trait(s) at this level? For type-1 agent: Have the traits at this level jointly evolved to fulfil a common goal?	Tinkered, engineered phenotypes (e.g., established by optimality models) *at the focal level,* characterized by the presence of trans-temporally accumulated adaptations. Necessary to discover evidence of a mechanism that makes accumulation possible	Kin selection, optimality modelling, heritability and adaptationist general modelling in multilevel genetics, multilevel modelling of the Wrightian school	adaptationist tradition in evolutionary thinking/genetics evolutionary change tradition in evolutionary modelling (sometimes)

| Replicator/reproducer/ reconstitutor | What type of object gets inherited? What mechanism ensures transmission of the phenotype, especially the capacity to develop, at the focal or lower level?/How frequent is the reappearance of the phenotype across generations? | Molecular evidence, evidence from evolutionary developmental biology. Evidence of mechanisms of heritability at the higher level | Mechanistic modelling; molecular biology; heritability for multilevel selection modelling, genomic and phylogenetic surveys | Reduction and emergent approaches to selection and evolution; evolutionary change and adaptationist approaches. ETI context |

Summary of the tripartite version of the DP about units, including the types of questions it captures, the type of evidence required for an affirmative answer to these questions, the type of modelling practices it uses, and the predominant research context where these questions are asked.

Three clarifications about tripartite version of the DP are in order. Firstly, while in the early versions of the tripartite framework the replicator came to be associated with certain pieces of DNA (possibly given the fact that the concept is itself an abstraction from DNA research, at least in Dawkins 1976; see Godfrey-Smith 2009), this association is by far a contingent one. The replicator is an epistemic or abstract object that captures a multiply realisable causal concept playing a functional role in the process of selection (and which we will later complement with the introduction of the reproducer and the reconstitutor). DNA, in contrast, is a possible realizer (among many others) of this crucial reproductive role (Veigl et al. 2022). In fact, the version of the tripartite framework we advocate in this Element acknowledges the role of the replicator/reproducer/reconstitutor in any units of selection questions concerning the process of evolution by selection. Which entities are relevant units in such a process *depends completely on the research question(s) pursued by the contemporary biologists within their research programmes* (Section 4).

Secondly, nothing in the tripartite framework we have presented impedes the possibility that one object is simultaneously an interactor *and* a replicator – or an interactor and a manifestor. Genes causing segregation distortion are a canonical example of an object that fulfils the interactor and replicator roles simultaneously, but not the only case (see Brandon 1988; Hull 1988b; Lloyd 1986, p. 398). The distinction between interactors and replicators appeals exclusively to the functional roles (phenotype-fitness vs. connection/heritability) that some objects must adopt in the process of natural selection, to be properly characterized as such, but not to the specific and multiple material realizers of these roles. The necessity of distinguishing these two functional roles historically derived from the perception of a confusion in the cases where different objects stood for each of the roles, as we have shown. But it is a mistake, unfortunately repeated in the literature, to infer from those cases that the two functional roles *always* need to be attributed to two different objects, where one of them – the replicator – necessarily corresponds to DNA.[19]

Finally, we have shown how the tripartite schema translates the recipe approach into different sets of research questions that can be asked (sometimes jointly, sometimes separately) in doing research about the units. These and related questions and their combinations are the basis of research about the units of selection in evolutionary biology. Distinguishing them is required; once the research questions are distinguished, we can distinguish the different types of evidence that are required to answer those distinct questions. This is all

---

[19] Although note that Dawkins never made this error himself, who perfectly knew that the replicator represented an abstract object.

essential to avoid crosstalk between different parties working on the different projects addressing units of selection. In this sense, we demonstrate the value of the tripartite version of the DP as an *epistemological and methodological* framework of research questions to study the different type of functional roles or subprocesses that the action of natural selection in different contexts and under different circumstances can give rise to.

4 The Framework of the Evolutionary Transitions in Individuality: A Challenge to the Disambiguating Project

The early versions of the bipartite and tripartite frameworks of the DP we have presented so far take for granted the existence of the function of replication – even though only some assumed the function to be played by a concrete entity or process (Lewontin 1970; Hull 1980) whereas others conceived replication fully abstractly (Lloyd 1988/1994; Griesemer 2000a). As such, it could be understood as taking the existence of a replication hierarchy for granted, and it is not a project that was in principle conceived to explain the origins of such hierarchy. To put it another way, the DP was conceived to frame the problems of how certain traits evolved, or how certain organisms and their populations trans-temporally accumulated certain designed or engineering adaptations, and to do so, it assumed that the replicators at one level reproduce independently from one another. While the product-of-selection evolution of some traits at one level does not require the joint replication of the replicators at that level – lower-level replicators simply need to replicate differentially – it seems that the same is not necessarily correct for the trans-temporal accumulation of adaptations at one level – which minimally requires the accumulation of these adaptations in one lineage, and hence a certain degree of joint replication *at that level*.[20] But in the way that we have introduced the DP, and how it used to work in the evolutionary change and adaptationist schools before the 1990s, the existence of the hierarchy of replicators is evolutionarily presupposed, rather than evolutionarily explained (Griesemer 2000a, 2000b). Ontologically, this can be expressed by saying that replication was taken as a primitive in early formulations of the DP, whereas the ETI shows that the origin of replication across reproductive levels is also a trait that requires to be explained adaptively, as a result of accumulation of design. Note therefore that the type of unit that ETI research is interested in is a manifestor/type-1 agent *with*

---

[20] But see Wade (2016) for a discussion elaborating the complexity of this claim; Stencel and Wloch-Salamon (2018) and Papale (2020) for conceptual evidence that the evolution of manifestors/type-1 agents may not always require joint reproduction at the focal level, and Suárez and Triviño (2020) for a case of evolution of manifestors/type-1 agents at one level without known collective reproduction at the level. Note, in any case, that what really matters from the ETI perspective is the evolution of manifestors/type-1 agents *with respect to their reproduction*, rather than with respect to any other type of optimal design.

respect to its own reproduction. Or, to put it differently, the type of designed trait that ETI researchers believe requires an adaptive explanation is the reproduction at the focal level of the higher-level unit. We will from now on use the term 'sequestration' to refer to the multiply realizable processes by which independent replicators start replicating together, as a single unit of replication.

Okasha (2006) refers to the way of presenting the problem of the units before the introduction of the ETI as *synchronic*. But note that the original formulation of the DP is only synchronic with respect to the emergence of different reproductive regimes, or levels. It is, in contrast, not happening all at the same time with respect to the emergence of design or engineering adaptations, which are assumed to appear diachronically as conveyed by their label *trans-temporally accumulated*. But we agree with him that a key evolutionary question not addressed in the anatomy of meanings of the expression 'units of selection' originally distinguished under the tripartite framework concerns what he calls the *diachronic perspective*, which we would more precisely characterize as the question concerning evolution of different modes of *replication*, that is, diachrony with respect to the emergence of the reproductive hierarchy regime or levels.[21]

The reason why the question about the origin of the reproductive hierarchy had been ignored in early work on the units resides in the functional role attributed to the replicator in the first versions of the DP. In the original formulations of the bipartite framework, the simultaneous presence of interactors *and* replicators is considered necessary and sufficient to infer the existence of a process of selection (Section 2). Additionally, the tripartite approach did not originally replace the role of the replicator in the earliest versions by any other type of reproducing unit.[22] This lack may induce one to believe, incorrectly, that even though natural selection may play a role now in the evolution of different objects in the biological hierarchy in virtue of the functional roles that these objects acquire – interactors, replicators, manifestors/type-1 agents – it has played no role in the very origin of these functional roles – especially the role occupied by the replicator, but also the role of the manifestor/type-1 agent *with respect to its own reproduction* – and hence on the origin of different objects in the reproductive hierarchy itself (Okasha 2006; Godfrey-Smith 2009). The project of the ETI proposes a diachronic approach that also aims to explain the origin of the reproductive hierarchy (Maynard Smith & Szathmáry 1995). That is, it concerned an entirely different research question: what is the *evolutionary origin* of the distinct functional roles assumed in any DP analysis, either in a given case, or generally?

---

[21] In what follows, the reader should understand *synchronic* and *diachronic* as referred exclusively to the evolution of the reproductive hierarchy, or the emergence of new levels *of reproduction.*

[22] Note that while this may be true of the earliest versions presented in Lloyd (1988/1994, 1992), it is not the case for most of the later versions she presented (Lloyd 1994, 2015, 2023).

Evolutionary transitions in individuality is built upon two key core Adaptationist assumptions: (1) the origin of objects in the reproductive hierarchy needs to be explained evolutionarily and adaptively, as a result of design, rather than taken for granted; (2) natural selection must have played a role in the origin *of the functional roles that these objects can acquire*, most especially of the role occupied by replicators.

Given that it might possibly be true that the ETI project poses a challenge to the tripartite Framework, several authors have interpreted it as a reason to abandon the latter view, and go back to the recipe approach as the basic description of the units of selection (Okasha 2006; Godfrey-Smith 2007, 2009; Bourrat 2021). But, as noted previously in Sections 2 and 3, *this characterization misconceives the Tripartite Framework, as well as the more general DP about units, for it fails to appreciate both the details and motivation of the general project, as well as the specificity of the Tripartite version of DP.*

This section traces the origins of the misconceptions about the DP, which we summarize in Table 2 to guide the readers. We show that the ETI project does not really invalidate the DP, or the tripartite version of it, but rather suggests that questions about a relatively new unit – the *reproducer* or *Darwinian individual* – should be incorporated within the framework. Additionally, we show that the type of questions asked under the ETI project can be interpreted in the light of the tripartite version of the DP, as a combination of the different questions isolated under the DP, suggesting its contemporary validity. Finally, we show that the current questions asked under the ETI project do not constitute a profitable research avenue to conceptualize the debates about units, *if these questions are taken as the only type of legitimate questions to be asked about units*, as is suggested by some of those working under the ETI framework.[23]

4.1 The Project of the Evolutionary Transitions as a Challenge to the Polysemy

Buss (1987) is probably the first author to show preoccupation for the problem of the origin of the reproductive hierarchy. He writes:

> Numbered among [a history of transitions between different units of selection] . . . must be the origin of self-replicating molecules, the association

---

[23] Note that we focus on the research on ETI starting in the 1990s and flourishing during the first decade of the twenty-first century, in which the central question in the ETI project was to understand the origin of reproduction across levels. We are conscious that the tendency has changed after 2010, in which questions about the evolution of other forms of individuality (metabolic, e.g., O'Malley & Powell 2016), or the evolution of different inheritance systems in general (e.g., Szathmáry 2015), also joined the list of the Major Transitions and thus expanded the events to be included. We leave the analysis of these cases out of this work.

Table 2 Summary of the main misconceptions about the tripartite version of the DP we advocate here

Misconception of the DP	Reality of the framework
(1) The cohesiveness of interactors is an evolved property that requires explanation, so the interactor concept is inadequate (insufficiency claim)	It depends. The 'cohesiveness' of interactors can be misleading. It primarily refers to the *unified response to the environment* of the interacting entity at that level whose effects then *spread down* to their replicators or reproducers. This 'cohesiveness' may be an evolved property, but it is not necessarily an evolved property in the sense of a product resulting from a process of trans-temporal accumulation of adaptations at the level where interaction occurs. However, sometimes the cohesiveness is higher, that is, it results from a process of trans-temporal accumulation of adaptations at the level. In the latter case, the concept of 'manifestor of adaptation' or 'type-1 agent' is preferred to avoid confounding questions concerning highly evolved engineering trans-temporally maintained adaptations (optimization questions, characteristic of the adaptationist school of evolution) with questions concerning where the process of selection is acting at a specific point in time on entities at a level (interactor questions, characteristic of both the adaptationist and evolutionary change school of evolution)

(2) The concept of the interactor is imprecise, so it should be abandoned or reformulated

The concept of the interactor is one of the most precise concepts in the debates about the units of selection, and many researchers have tried to make the concept more precise by appealing to different methods of detection (context-independency, additivity, screening off, common fate), many of whose applications yield the same results. Overall, determining the existence of an interactor at one level requires the combination of empirical evidence with technical statistical tools (contextual analysis, or other types of regression analysis) that ensure that interaction at that level is genuine and not a cross-product of selection at different levels. Different accounts of the interactor, possibly even arising from different schools in evolution, where the interactor and manifestor have been conflated (Misconception 1), may disagree about which statistical tools are optimal in different selective environments; they may also disagree about whether a single, general tool, rather than a plurality of tools, is preferable. But the concept of the interactor remains valid, as it is used in the biological literature

True only of early versions of the tripartite framework, but not of recent versions or the version we advocate here. The necessity of looking for replicators somewhere in the system is contingent

(3) The interactor, replicator, manifestor/type-1 agent framework is inadequate because the presence of replicators somewhere in the system is always required

Table 2 (cont.)

Misconception of the DP	Reality of the framework
(lack of necessity claim)	upon the trait of biological interest. For some traits, the presence of reproducers is the only requirement. That's why the framework includes the interactor, replicator/reproducer/reconstitutor, and manifestor/type-1 agent. The introduction of the 'reproducer' into units analysis dates back to 1994, and it was primarily introduced in the context of development and the ETI project, as this high-lighted the problem with the narrower functional role attributed to the replicator in the original formulations. The introduction of the reconstitutor is recent, to account for forms of phenotypic reappearance over generations without material overlap or formal causal influence from the parents on the offspring.
	Note that the necessity of replicators/reproducers/reconstitutors is required to distinguish cases where evolution by natural selection can be acting from cases of phenotypic plasticity in the process of growth
(4) An object cannot be simultaneously an interactor, replicator/ reproducer/reconstitutor, and manifestor/type-1 agent, but the ETI	Under the tripartite framework, an object or entity has always been framed to be *able to play* the three roles simultaneously for certain traits. But it is not *necessary* that it does so. Hence, nothing in the

project shows that some objects simultaneously play these three roles

formulation of the tripartite framework blocks the possibility that one single object plays the three roles for a specific trait. On the contrary, the tripartite analysis explicitly allows for such cases which may involve complex evolutionary processes at multiple levels

(5) Contemporary biological debates about units only concern ETI, so the necessity of having a tripartite project disappears, because the concept has become univocal

Contemporary biological research still includes research projects about interactors, manifestors, replicators and reproducers, and combinations of these functional roles, and while these questions are sometimes asked jointly (as in the case of the study of the ETI), this is not one research question about the units, but a set of theoretically distinct and complementary questions. Failing to appreciate this has led to serious confusions of autonomously replicating molecules into self-replicating complexes, the incorporation of such complexes into cells, the establishment of a multigenomic cell via the incorporation of autonomously replicating organelles, and, with the evolution of sexuality, the origin of species. (Buss 1987, p. 171)

And he continues:

> The organization of units of selection . . . bears these two features in common. First, each lower-level unit is selected by features of the organization of the higher unit . . . . In each case, the lower-level unit may replicate, but only within bounds set by the influence of this replication on the higher-unit's effectiveness in its interaction with the external environment. Second, the unit itself is stable . . . . Each unit has manifestly resisted significant perturbation over geological timescales. (Buss 1987, pp. 171–2)

There is an inherent tension underlying Buss' ideas, which make him a clear precursor of the ETI research project, but not exactly the one generating the programme. On the one hand, Buss shows a clear preoccupation with how different forms of organization and modes of replication evolved. In a sense, this derives from his interest in incorporating development in evolution, expressed throughout his Element. On the other hand, he is not clear about whether natural selection plays any specific role in creating these different levels of organization, or whether it only starts acting once the level has become organized *enough*, that is, once interactors and replicators are clearly identifiable at one level, and become transgenerationally recursive at this level to allow the emergence of tinkering/engineering or trans-temporally accumulated adaptations (see esp. pp. 183–8; Fontana & Buss 1994). Another notable element in Buss's presentation is his merging of three functional roles into the one story. His attitude towards the relationship between the ETI and natural selection functional processes is similar to recent ideas expressed by Ellen Clarke (2013, 2014), even though she is slightly more conscious of the potential role of natural selection *at certain early stages* in this process.

Maynard Smith and Szathmáry (1995) give the clearest articulation of the project of the ETI, and its connection to the debates about the units of selection. They present the project as a quest for the evolution of complexity and concentrate it on the study of the ways in which information is stored and transmitted between generations. They assert:

> One feature is common to many of the transitions: entities that were capable of independent replication before the transition can replicate only as part of a larger whole after it. . . .
> Given this common feature of the major [evolutionary] transitions, there is a common question we can ask of them. Why did not natural selection, acting

on entities at the lower level (replicating molecules, free-living prokaryotes, asexual protists, single cells, individual organisms), disrupt integration at the higher level (chromosomes, eukaryotic cells, sexual species, multicellular organisms, societies)? (pp. 6–7)

And they declare their background assumption about how the ETI project must be framed:

The transitions must be explained in terms of immediate selective advantage to individual replicators: we are committed to the gene-centred approach outlined by Williams (1966), and made still more explicit by Dawkins (1976). (p. 8)

Note that this way of presenting the project, and the key dynamical and structuring problem that motivates it (why selection below did not disrupt selection above), is grounded on what Lloyd and Wade (2019) call a cheater-point-of-view, which they argue is characteristic of the adaptationist school (Section 2). But even if this is so, it is true that the project of the ETI could be – and has been – interpreted as one that poses a fundamental tension in the functionalist (bipartite or tripartite) approaches to the levels of selection.

On an ETI view, the functionalist tripartite approach presupposes the existence of objects with 'evolved' properties: 'cohesive interactors' [manifestor/type-1 agents],[24] which are also replicators with a high degree of copy-fidelity, longevity, and fecundity. But the ETI approach places its interest in the *diachronic* dimension with respect to reproduction. That is, how objects with these 'evolved' reproductive properties originated, and the role that natural selection must have played in shaping or even tinkering/engineering objects with these very reproductive properties (Griesemer 2000a, 2003). This creates the appearance of a problem, though, given that the existence of objects that play the functional roles of 'cohesive interactors' [manifestor/type-1 agent] and replicators seems insufficient to explain all selection processes, particularly, the origin of the objects that play these functional roles (Misconception 1, Table 2). We will refer to this criticism as the 'insufficiency claim'. Okasha has succinctly expressed what researchers working on the ETI project perceive as the key problem with the received bipartite/tripartite frameworks:

---

[24] The problem is that some researchers may think that the type of cohesiveness that interactors need to have concerns the evolution of designed or trans-temporally accumulated adaptations that generate an emergent regime of reproduction. But this conflates a manifestor/type-1 agent requirement about cohesiveness with an interactor requirement about cohesiveness, which is completely different and perfectly compatible with the ETI project. See Misconception 1, Table 2. For the rest of this section, the reader must assume that when we use 'cohesive interactors', we mean *manifestor/type-1 agents*, whereas when we use 'non-cohesive interactors', we mean *interactors*. We will indicate this in brackets.

Since the levels-of-selection debate now encompasses questions about the
origin of the biological hierarchy, not just the evolution of adaptations at
preexisting hierarchical levels, an abstract characterization of Darwinian
principles cannot refer to highly evolved features, of either organisms or
genetic systems, on pain of an inevitable loss of generality. Characterizations
in terms of "high-fidelity replication" and "cohesiveness" fall foul of this
constraint. (Okasha 2006, p. 14)

Note what is at stake here. The original functionalist bipartite approach to the
units of selection as introduced by Dawkins and Hull was devised to study how
certain phenotype-fitness connections at a level affect the pattern of heritability
at the same, or at a lower level. A general presupposition in Hull's version of the
framework was that the level where interaction occurs would get tinkered or
engineered – accumulate adaptations trans-temporally – as a result of the action
of natural selection.[25] But its existence as a level that can get tinkered was
presumed. A similar presupposition, but with respect to high-fidelity copying,
was made about replicators in both Hull's and Dawkins' accounts. The existence of replicators was taken for granted and not considered to require an
evolutionary explanation.

The ETI project, in contrast, is devised by assuming that *the* key and –
according to some – *only* question in the units of selection consists in understanding the tinkering or engineering process that generates the levels that can
evolve further engineering or transgenerationally accumulated adaptations *by
sequestrating the independent replication of previously replicating entities*.
This requires asking a question about how what they call 'cohesive interactors'
[manifestor/type-1 agent], and high-fidelity copying replicators, evolved in the
first place. Buss had appreciated only half of the problem – the origin of the
reproductive hierarchy – but had failed to appreciate the role of natural selection
in generating the type of group adaptations that ensure the transition, that is, the
sequestration of the replicators. When replication is sequestered, it is almost
ensured that the new 'collectively replicating' unit, which is both a reproducer
and a manifestor/type-1 agent – which has to be an interactor as well – can
evolve genuine 'benefits', and 'altruistic' traits in the restricted sense that
Williams and Maynard Smith had used the terms in the 1960s (Okasha 2003).

But note that the ETI project, and the question it raises, constitutes only *one
dimension* of the units of selection debates (Misconception 5, Table 2): the one
that reduces the questions about the units of selection to the question about the

---

[25] Note that the presupposition regarding tinkering does not directly apply to the Tripartite
Framework, because the concepts of *interactor* and *manifestor/type-1 agent* are kept apart. It
does only insofar as this tinkering refers to the regime of reproduction, in which case the
reproducer needs to be added to the Framework, given the insufficiency of the replicator and
interactor to account for that type of tinkering (Reality 3, Table 1).

evolution of manifestors of adaptation or type-1 agents via sequestration of independent replicators. Or, in other words, to the question concerning the conditions under which 'group selection for group benefit' can evolve with respect to a regime of reproduction – a question which simultaneously involves other questions, for example a question about interactors, but which are framed under the same research project.

Furthermore, the criticism of the cohesiveness of the interactor (Misconception 1 Table 1) conflates genuine questions about the role of the interactor and its relations with its environment, with questions about the manifestor/type-1 agent, which as we showed had been conflated in bipartite approaches to the DP, such as Hull's or Sober and Wilson's. But the concept of the interactor is not exclusively fulfilled by objects that are simultaneously manifestors/type-1 agents. Some objects may play the role of an interactor also in those stages where engineering or trans-temporally accumulated adaptations have not evolved – or even will never evolve – at the focal level. An interactor at the higher level exists whenever there are joint interactions between the entities at the lower level, which are phenotypically expressed at the higher level in a way that causally biases the expected replication or reproduction of the entities at the same or at the lower level (Table 1).

Hence, if *cohesiveness* were conceived as an evolved property of the engineering/trans-temporally accumulated type, the insufficiency claim would simply assert that the ETI project is studying how what they see as 'non-cohesive interactors' – entities that have not evolved trans-temporally accumulated adaptations, our interactors – at the higher level, composed of independent replicators, *and* what they call 'cohesive interactors' [manifestor/type-1 agent] at the lower level, may evolve into a joint replicator or reproducer at the higher level that is simultaneously a 'cohesive interactor' [manifestor/type-1 agent] *with respect to its own reproduction* at the higher level (Misconception 1, Table 2). Note that, if this is the real dispute, the insufficiency claim merely involves an equivocation on the term 'cohesiveness' between interactor and manifestor/type-1 agent cohesiveness, rather than a substantial discussion about natural selection or the validity of the tripartite version of the DP. We contend that, if one properly conceives the tripartite version of the DP, including especially the tripartite version we advocate here, the insufficiency claim constitutes such an equivocation.

The introduction of the ETI framework has another key consequence for the debate about the polysemy of the expression 'units of selection'. This is one which, in contrast with the other criticisms that we believe are misguided, we contend to be essentially correct, thus requiring a subtle revision in the concepts isolated under the tripartite version of the DP we advocate in this work

(Misconception 3, Table 1). The ETI project originated by presupposing the primacy of the replicator, in the form of the genic approach, as the quote by Maynard Smith and Szathmáry reflects. Yet it puts an essential tension on the concept of the replicator itself, as well as its necessity in describing the process of natural selection. This tension can be encapsulated as follows: if new levels of selection are created by generating 'collective replicators' that in the process *gradually* lose their capacity for independent replication, then the same *graduality* might have been a characteristic of the evolution of the first quasi-replicating molecules. In other words, the first 'replicators' must have been very different from the highly evolved – with trans-temporally accumulated adaptations in relation to their own reproduction – coding replicators that we know now. But then, the replicator, as defined in the bipartite and original versions of the tripartite analyses, is unnecessary for the process of natural selection, a mere contingent result of how evolution has affected some biological objects (see also Okasha 2006; Godfrey-Smith 2009; Bourrat 2021). We will refer to this criticism as the 'lack of necessity' claim.

4.1.1 The Project of the ETI and the Recipe Approach to Natural Selection

Maynard Smith and Szathmáry's ETI project concerns primarily replication, and how this functional role can shift between objects or levels of the biological hierarchy. In a sense, the project may be taken as leading to a replacement of the functionalist formulations of the DP, because the object that occupies the functional role of the interactor loses importance on its own, and only gains it insofar as it is simultaneously either a reproducing object, a manifestor of adaptation/type-1 agent with respect to its own replication, or both. This does not necessarily entail that interactors become irrelevant in the ETI project: fitness-affecting interactions and non-additive effects between independent replicators are fundamental in the genetic models that study these transitions. But the inner logic of the research questions posed by the ETI framework makes them only relevant *insofar as they can model the type of mechanisms or specific conditions (or constraints) by which currently known collective replicators could have evolved.*

The starting empirical observations in the study of evolutionary transitions are always from:

1. one object at the higher level that is known to be *simultaneously* an interactor, a 'replicator' (rather, a 'reproducer', see Section 4.2), and a manifestor of adaptation/type-1 agent;
2. one or a few objects at the lower level that are known to have been simultaneously interactors, replicators, and manifestors of adaptation/type-1 agents, and *are known to have* (at least partially) *lost* one or more of these roles.

This combination of the adaptationist logical structure with the empirical observations about the ETI gives rise to a specific formulation of their research questions concerning the evolution of different reproductive regimes. As we said in Section 3, a typical Adaptationist research problem would encompass two research questions:

1. Identification of a higher-level trait that is optimal at that level but seems to go against the selective interest of the entities at the lower level (genes, organisms, independent lineages), making the trait unlikely to evolve;
2. Imagination of a selective scenario where this trait could have been evolved against the odds.

In the study of ETI, the optimal higher-level trait is the *joint reproduction* of previously independent reproducers or replicators. This trait would have evolved against the tendency of lower-level reproducers or replicators *to reproduce or replicate independently* from each other.

Given this way of formulating the research questions, the adaptationist school taking on ETI issues can generate a research programme that, instead of taking the existence of replication or reproduction as a primitive and aiming to explain the evolution of some types of design (manifestors/type-1 agents), aims to explain the very origins of the reproductive hierarchy itself as the result of a cumulative process of tinkered/engineered adaptation at the reproductive level.[26]

Note the centrality of the question about the manifestor/type-1 agent in these debates, which reveals the adaptationist and cheater-point-of view origins of the project (Lloyd & Wade 2019). The reproductive hierarchy is conceived as a result of an engineering process of trans-temporally accumulating adaptation on different and successive units, which affects their modes of replication. This leads Maynard Smith and Szathmáry to the explicit rejection of several cases as cases of evolutionary transitions – which is reasonable, given the project's nature in understanding the evolution of forms of replication – but also to the explicit a priori rejection of some objects as units of selection, which masks a confusion about the polysemic character of the expression. For example, they say:

---

[26] An interesting historical issue concerns why questions about the origin of replication or about the origin of different regimes of reproduction were not considered in evolutionary research before the ETI project was introduced. A possibility concerns the division of labour between different research traditions, with questions about the origin of replication assumed to be part of the field studying the origins of life, rather than its evolution. But maybe the development of new technical tools may have been somehow involved in this change, for example, superfast genomic sequencing of bacterial and other unicellular entities that produced a new 'web' of life, at the root of the tree of life (Doolittle & Bapteste 2007). This new web of life posed challenges to ordinary views of the possible entities that are subjected to natural selection, especially because of the interactions and gene exchange among the low-level entities. We leave the study of this question for further work.

> It might be asked why we do not include the origin of ecosystems in our list of transitions. There are two reasons. First, in temporal terms an ecosystem is not the final stage in a series: ecosystems are as old as replicating molecules. Second, ecosystems are not individuals, separated from others, whereas the other stages we have listed (including sexual species, and insect colonies) do have a degree of individuality, and separateness from other entities of the same kind. For this reason, ecosystems cannot be units of selection. (Maynard Smith & Szathmáry 1995, p. 7)

It is not surprising that ecosystems are excluded from their list, while they were not surprisingly included in the original list of Lewontin:

> At yet higher levels, the species and the community, natural selection obviously must occur. Species evolve to survive in a certain environmental range, and if the environment should suddenly change, some species will become extinct but others will survive. *The same is true of communities whose stability of composition depends upon the interaction among their constituent species.* If the rate of environmental change is slow enough, natural selection operating within species will allow the phylogenetic line to survive even though the species may slowly transform. Yet more than 99 per cent of all phyla that have ever existed are extinct. *On the geological scale of time, then, there is selection between phyla for those best able to survive even the slowly changing environment.* At times environments shift more rapidly, as in the Pleistocene, so that extinction rates are very high. More recently, environmental change has been very rapid so that species as units are being selected quite often. Pests, weeds, and commensals with man, like Rattus and Mus, are all examples of species that, formerly rare, were accidentally preadapted to human environments and especially to the monoculture of certain crop plants and animals. These preadapted species have undergone an explosive increase as compared to the extinction or near extinction of those wild relatives that were unable to escape from the old habitats. When the habitats were destroyed, so were the species. Thus man, by his rapid transformation of habitats, creates the conditions for efficient natural selection of species *and species relationships*, on a scale never before existent in evolution. (1970, pp. 15–6, emphasis added)

We interpret this disagreement in terms of a difference in focus: while Lewontin's recipe approach can legitimately be interpreted in terms of different functionalist versions of the DP, as we have demonstrated (Figure 1, Figure 2), the same is not true for the ETI projects. ETI projects only focus on the specific type of research questions surrounding reproduction, and formulate them as questions about the origin of design or engineering adaptation and selective interaction at the reproductive level.

The most important consequence of this biological/philosophical interest in the ETI and its lack of coherence with certain topics previously included in debates about units – for our goal in this Element – is that several philosophical researchers who started to develop an interest in studying the foundations of the ETI project ultimately generated some confusions in contemporary units of selection debates. These researchers took the ETI project as a basis for developing a new approach to thinking about the units that allegedly would go beyond the DP. We believe that these confusions would not have arrived had the DP been kept at the centre of the debates about units and used as the correct tool to interpret the ETI project – for example by developing revised new concepts of 'units' – rather than using the ETI as the tool to reinterpret the whole unit of selection debates.

The most important mistake of those who took the ETI project as a basis for rethinking and rejecting the whole DP about units lies in assuming that *all debates about units are ultimately debates about the ETI*. This leads to the inadequate reduction of a pluralistic field of research, where different meanings for the expression 'units of selection' are used, to a monist one, where one single question – about the evolution of reproduction – is asked (Misconception 5, Table 1). While ETI questions are legitimate, especially under the adaptationist framework of evolutionary thinking, the reduction of all debates about units to debates about the ETI is not, as it leads to an eliminativist version of the UP about units which squares poorly with today's biology.

4.2 Taking the Evolutionary Transitions in Individuality Challenge Seriously

The most pressing aspect of the ETI challenge to different versions of the DP concerns the necessity and sufficiency of the replicator in discourses about natural selection. There are two possible responses to the challenge: on the one hand, one can keep the DP in any of its versions, replacing the concept of the replicator with an alternative one that covers the replicator but also all the intermediate events during the transitions; on the other, one can break radically with the DP and look for a unique meaning of the expression 'units of selection' that encompasses all the previous meanings. Griesemer adopts the first approach (Section 4.2.1), whereas Godfrey-Smith attempts the second (Section 4.2.2).

4.2.1 Introducing the Reproducer

Griesemer proposes to replace the concept of the 'replicator' with the concept of the 'reproducer'. While the replicator focuses on the fidelity of copying mechanisms in certain genetic elements, the *reproducer* instead focuses on the material transference of genetic and other matter from generation to generation (Griesemer 2000a, 2000b; see Forsdyke 2010). Griesemer argues that a reproducer is the entity that engages in the process of reproduction, which he defines as follows:

> The process of reproduction can be analyzed as multiplication of material overlapping propagules that confer the capacity to develop, specified in terms of the minimum notion of development as acquisition of the capacity to reproduce . . .. The interlocking of developmental and reproductive capacities is recursive. The realization of a reproduction process entails the realization of a developmental process. The realization of development entails reproduction. The recursion bottoms outs in a condition of null development. (Griesemer 2000c, p. 74)

On this approach, thinking in terms of reproducers incorporates development into heredity and the evolutionary process. It also allows for both epigenetic and genetic inheritance to be dealt with within the same framework. The reproducer plays a central role, along with a hierarchy of interactors, in work on the units of evolutionary transition or ETI. This can be clearly seen in Griesemer's (2000c) serious criticism of the 'replicator first' approach advocated by Maynard Smith and Szathmáry (1995, p. 8):

> Viewing the relation between replication and reproduction as Szathmáry and Maynard Smith have is to adopt Dawkins' gene's eye view, in which replicators are rare and the subjects of a fundamental process (replicator selection) while interactors (reproducers) are common, hierarchically organized subjects of a derivative process (vehicle selection). I argue for a more radical perspective, in which replicators are viewed as a special class of reproducers whose development is deeply dependent on many higher levels of reproductive organization. Replicators are deeply dependent because successive evolutionary transitions have altered their mode of transmission and information storage – i.e., their mode of *development* – several times over. (2000c, p. 77)

Interpreting that sentence may be complicated, as Griesemer seems to be equating the hierarchy of interactors to the hierarchy of reproducers; in other words, he seems to be talking as if the hierarchy of reproducers would have been introduced to replace the hierarchy of interactors, as well as the very concept of the replicator. But in this paragraph he is specifically talking about how to understand the ETI project in a way that generates a more interesting heuristic than the one he thinks the original Maynard Smith and Szathmáry's coinage in terms of 'replicators' generates. Interpreting Griesemer as if he were rejecting the polysemic meaning of the units of selection *tout court* is simply a mistake. In Griesemer's view, there is nothing essentially mistaken in the functional analysis of the levels that divide the debate into several debates depending on the questions that are being addressed, provided the latter is adequately contextualized. He says:

[A]s long as evolutionary theory concerns the function of *contemporary* units at *fixed* levels of the biological hierarchy . . . the functionalist approach may be adequate to its intended task. However, if a philosophy of units is to address problems going beyond this scope – for example to problems of evolutionary *transition* . . . then a different approach is needed. (Griesemer 2003, p. 174)

The 'reproducer' concept, which as we have shown incorporates the notion of development into the treatment of units and levels of selection, is a step towards meeting the goal of addressing such evolutionary transitions, and, pertaining to evolutionary time, Griesemer writes: 'the dependency of formerly independent replicators on the 'replication' of the wholes – the basis for the definition of evolutionary transition . . . is a *developmental* dependency that should be incorporated into the analysis of units' (Griesemer 2000c, p. 75).

Note that Griesemer says 'incorporated into' and not 'replaced'. For his basic idea, as we interpret it according to the evidence we have shown, is that the ETI project shows that the functional role assigned to the replicator, as the *unique* marker of heritability, had been inadequately captured by Dawkins, Hull, and Maynard Smith and Szathmáry, because it forgets development and, hence, the possibility that there are other channels of heredity (for recent developments, see Merlin 2017; Merlin & Riboli-Sasco 2017; Veigl et al. 2022). But this criticism does not extend to any of the other functional roles. Interactors and manifestors of adaptation would still be valuable tools in analyzing debates about the units because they pick out functional roles that are experimentally studied (Goodnight & Stevens 1997; Wade 2016).

Overall, we think Griesemer's diagnosis about the limitations of the concept of the replicator is correct, and hence we adopt his idea that many (but not all) research projects framed as projects about the 'units' concern the reproducer as the unit of inheritance (Reality 3, Table 2). In doing so, we avoid the presupposition that evolution by natural selection occurs only when there are evolved coding mechanisms of replication, which was implicit to the concept of replicators. In the specific case of the ETI, the reproducer concept allows us to separate two elements (see also Griesemer 2000c, p. 77):

1. The basic development involved in the origin of a new biological/reproductive level [interactor, reproducer questions]. This would correspond to the part of the adaptationist research question concerned with imagining a selective scenario demonstrating that there is a trait that would seem to have evolved against the odds, primarily as a product-of-selection adaptation.

2. The later evolution of sophisticated developmental mechanisms for the 'stabilization, maintenance and perfection of a new level of reproduction' [interactor, manifestor/type-1 agent questions]. This would correspond to

the part of the adaptationist research question concerned with the identification of a higher-level trait (joint reproduction) that is optimal at that level but seems to go against the interest of the entities at the lower level.

These two processes constitute our preferred way of reformulating the essence of the project of the ETI under the tripartite version of the DP we advocate in this Element. They include two of the three types of questions we introduced in Table 1, plus the fundamental question about the reproducer that Griesemer introduced – and the reconstitutor that Veigl et al. (2022) introduced later – to expand the replicator concept, which we accept the DP must include (Reality 3, Table 2).

4.2.1.1 The Reproducer and The Disambiguating Project about the Units

We would finally like to note an important reason why Griesemer introduced the reproducer concept, which we have not introduced yet. This connects to an issue we articulated in Section 4.1.1: namely, that all debates about ETI already presuppose the existence of a higher-level object which is simultaneously an interactor, reproducer, and manifestor/type-1 agent, as well as several objects in the lower level that once fulfilled the three roles, but do not embody them anymore. These questions can hardly be understood solely based on the interactor/replicator concepts, without introducing a manifestor/type-1 agent concept. But a futher concept that is flexible enough to accommodate precisely the *gradual character* of transitions – a concept that captures how new entities with the appropriate statistical features of phenotypic variance in fitness with respect to the reproductive mode appear (i.e., are interactors), and how these entities can evolve engineering or trans-temporally accumulated adaptations at their level of reproduction – is also required. The reproducer concept fulfils this latter role, in a sense that acquires special meaning in the transition between multilevel selection type-1 models (MLS1) and multilevel selection type-2 (MLS2) models (Heisler & Damuth 1987; Damuth & Heisler 1988; Lloyd 1988/1994; 2018, 2023; Okasha 2006).

In these types of evolutionary models, as wielded by biologists using the ETI approach, the higher level is *always* an interactor, whereas the lower level is an interactor and a reproducer only in MLS1 and gradually loses its independent reproducing capacity in MLS2, when they get 'transmitted' to the higher-level unit.[27] The loss of the reproducing capacity is gradual, and results from the evolution of engineering or trans-temporally accumulated adaptations in the higher level with respect to the process of reproduction.

---

[27] Note that this is an idealized story to capture how ETI works. But, obviously, complete suppression of selection at the lower level almost never occurs, and given recent discoveries about lateral gene transfer, the same may be true in the case of replication.

Under the ETI evolutionary picture, these engineering or trans-temporally accumulated adaptations at the higher level block the independent reproduction – hence the theorized opportunity for cheating, see Lloyd and Wade 2019 – of previously reproducing lower-level particles, to the extent that they become integrated into the higher-level unit. After the transition, the higher-level unit becomes an object that *simultaneously* fulfils the three functional roles. Once the transition is finished, the object at the higher level may evolve as an interactor for certain traits – favouring the differential replication of certain replicators within the reproducer that it simultaneously is – or as a manifestor/type-1 agent, accumulating new engineered adaptive traits over evolutionary time. Hence, after the transition, questions about the reproducer become mostly irrelevant (unless the object undergoes a new transition), because what the process of selection will do is either change the distribution of the extant traits – interactor questions – or engineer or tinker with the traits of the higher-level object – manifestor/type-1 agent questions, see Díaz (2017); Suárez (2019, pp. 162–72, 2021). So, we are again led into a synchronic perspective – with respect to reproduction – where the concept of the replicator, together with those of the interactor and the manifestor/type-1 agent, becomes again the most profitable avenue for channelling discussions about the units of selection.

Note however that this idiosyncratic and restrictive way of conceiving MLS1/MLS2 models is a feature of the ETI project, which cannot be generalized to every other project about the units, where MLS1 and MLS2 modelling can be applied simultaneously. In the end, these two forms of modelling selection processes are only two ways of modelling an evolutionary process, and they explain certain variances in the distribution of traits in a population (Heisler & Damuth 1987; Damuth & Heisler 1988; Jeler 2020).

4.2.2 Introducing the Darwinian Individual: Rejecting the Disambiguating Project

Godfrey-Smith's conception of the units of selection builds upon the ETI project, but in contrast to Griesemer, Godfrey-Smith introduces a new framework to replace completely the tripartite framework. In his view, the key problem with the interactor/replicator framework is the necessity of systematically finding replicators *somewhere* in the system to speak of a selection process (Misconception 3 Table 2).[28] He agrees that selection processes may

---

[28] Note that this is not the case if one accepts Griesemer's (2000a, 2000b, 2003) correction and conceives the DP as one that encompasses interactors, replicator or reproducers, and manifestors/type-1 agents, as this version accommodates the observation that heredity is sometimes

sometimes be fuelled by replicators, but he also contends that one should not forget that replicators are at most 'helpful tools' in a very special subset of cases where natural selection is acting. Yet,

> what is needed [for evolution by natural selection] is that the state of parents correlate with that of their offspring; parent and offspring must be predictive of each other to some extent, or more similar than unrelated individuals. . . . [W]eak parent–offspring similarity *when combined with selection* is often enough [for evolution by natural selection]. (Godfrey-Smith 2009, p. 34, emphasis added)

This is in fact one of the key consequences of the ETI project as introduced by Maynard Smith and Szathmáry (1995): namely, replicators cannot be seen as necessary for natural selection, because the first replicators must have been very bad at producing good copies of themselves. This conclusion is correct and has been part of the DP, as we have already explained (Section 2.1), and Godfrey-Smith agrees with it and takes it as the cornerstone of his project. But that's not the only consequence that he believes follows from the ETI project that inspires his view on the units of selection. Godfrey-Smith believes that in having gotten rid of the replicator as an essential concept in the analysis of units, his project must consist in finding a unique framework that unifies all questions about the units into a single unique question (Misconception 5, Table 1).

What is surprising, though, is that even his way of presenting the problem with the concept of replicator implicitly relies on two functional roles: on a distinction between parent–offspring similarity and the selection process itself, as our emphasis shows. He may argue that this is because he is relying on a distinction between *natural selection* and *evolution by natural selection* (or *response to selection*),[29] which do not have the same requirements because while selection per se does not require heritability, response to selection does insofar as biases in the hereditary pattern may counterbalance the trait-fitness connection. This, again, is correct, and probably the most innovative aspect in Godfrey-Smith's thinking is that he joins Griesemer in looking for an account that allows incorporating alternative pathways of heredity besides replicators (e.g., epigenetics) that can simply generate a small degree of parent–offspring similarity.

---

channelled by replicators and sometimes by reproducers. This option is taken by Lloyd (e.g., 1994, 2001, 2023) and we think it is correct.

[29] As we have already noted, detailing a possible course of natural selection that could have produced a particular believed-to-be-unlikely higher-selective higher-level trait is a standard project of the Adaptationist School; while tracking Evolutionary Change and its pathways or response to selection is a standard project of the Wrightian genetics or Evolutionary Change School of multilevel selection modeling, discussed especially in relation to the Tripartite Framework in Section 3.

However, in admitting that there is such a distinction between *selection* and *heritability* to be made, Godfrey-Smith must be implicitly admitting that in the study of evolution by natural selection there may be at least two types of research questions about distinct functions: questions about heredity – and the role of a replicator/reproducer/reconstitutor – and questions about selection per se – and the role of an interactor. Given Godfrey-Smith's way of framing and discussing the problem, this set of questions may also include a further question about the trans-temporal accumulation of adaptations from selection processes – the role of a manifestors/type-1 agents.[30] However, Godfrey-Smith does not pursue this analysis, probably because of his conviction that even if there are two questions, these must be addressed *at the same time* and by looking for one single object, which simultaneously fulfils at least to a very minimal extent all these roles (see esp. 2009, pp. 109–28) (Misconception 4, Table 1).

The next step in Godfrey-Smith's analysis consists in isolating the characteristics of *the* unit of selection. In his view, as the essence of the process of natural selection is the existence of parent–offspring similarity between the elements that constitute a Darwinian population, *the* unit of selection must be any element of a population of causally interacting elements where each of these elements has the capacity of forming parent–offspring lineages with at least a weak degree of parent–offspring similarity (i.e., individuals that have the capacity to reproduce, or reproducers, see Godfrey-Smith 2015, p. 10120; Griesemer 2000a, 2000b, 2000c). He refers to these elements as *Darwinian individuals*, and to populations composed of Darwinian individuals as *Darwinian populations*.[31]

Darwinian individuals are defined by their capacity of reproducing, but what really matters for Godfrey-Smith is how this capacity is mechanistically realized in different lineages. He distinguishes the types of Darwinian individuals, starting with *simple reproducers*. They are entities that reproduce with their own resources but nothing under their level (i.e., none of their parts) reproduces independently as well. A second type, *collective reproducers*, are entities that reproduce using their own resources but have lower-level parts that also reproduce independently. And finally, a third type called *scaffolded reproducers* are 'entities which get reproduced as part of the reproduction of some larger unit . . . or that are reproduced by some other entity'

---

[30] Note that this is an epistemological/methodological way of presenting the issue. For metaphysically inclined readers, the different questions may be considered to respond to the existence of different underlying properties: selection, heritability, designed adaptation.

[31] The expression 'Darwinian individual' was originally introduced by Gould and Lloyd (1999), although the concept they use, unlike Godfrey-Smith's whose concept conflated three functional roles of evolution-by-selection processes, was confined to the interactor.

(Godfrey-Smith 2009, p. 88). These three concepts are abstracted away, respectively, from asexual bacterial reproduction, multicellular organisms' reproduction, and viral reproduction.[32]

Out of these three reproductive modes, collective reproduction is the one that Godfrey-Smith develops more, as he thinks that is one of the key concepts needed to understand that Darwinian individuality is a graded concept (such that some individuals score higher, that is they are 'more Darwinian individuals' than others). In his account, collective reproduction can be identified whenever at least one of these three mechanisms can be identified in a collective of reproducing units: germ soma separation, overall integration, and/or the presence of a bottleneck. These three mechanisms generate a threefold structure in which collective reproducers can be scored according to their degree of Darwinian individuality: if one organismal lineage has evolved the three mechanisms, then it scores high and can be considered a paradigmatic Darwinian individual; if it lacks the three, then it is not a Darwinian individual; if it has any of them, but not all, then it will not be a paradigmatic Darwinian individual, but will still have a certain degree of Darwinian individuality.

Drawing on this conceptual characterization of '*the* question about the units', Godfrey-Smith argues that he can clearly distinguish those objects that are real units of selection from those that are not. Among the latter, he includes cases such as buffalo herds, or, in his later work, most symbiotic associations (Godfrey-Smith 2013, 2015; Booth 2014; Skillings 2016). Under Godfrey-Smith's lenses, these cases would be examples of ecological communities, or populations of independent individuals, where the organisms causally interact but do not give rise to parent–offspring lineages at the global level, so they cannot be considered real Darwinian populations (cf. Stencel 2016 for the view that some symbiotic associations can still be considered Darwinian populations even though they do not directly reproduce).

Furthermore, Godfrey-Smith will reject characterizing these communities as selective interactors, given that he perceives the concept imprecise and inadequate to capture (one of) the meaning(s) of 'unit of selection', insofar as it does not require any degree of inheritance at the focal level where interaction takes place (Godfrey-Smith 2011; see also Booth 2014) (Misconception 2, Table 1). Note that the necessity of some hereditary relations at the focal level plays such

---

[32] His characterization also includes formal reproduction, and his scaffolded reproduction would be a species of formal reproduction although the latter goes further. A case of formal reproduction that is not simultaneously scaffolded reproduction would be the reproduction of prions, in which a prion formally (but causally) influences a protein to become a prion. When this occurs, there is not material overlap at all between different generations of prions, but there is clearly a causal influence from the previous generation and hence the 'parents' can be identified.

a central role in Godfrey-Smith's conception plausibly because the evolution of reproduction – as opposed to the evolution of any other trait – constitutes the main issue at stake in the ETI debates about units (Section 4.1.1). Given that Godfrey-Smith builds his framework inspired by the work done under the ETI, it is understandable that he inherits such emphasis on reproduction of any given unit. But note that taking this specific requirement of ETI as a general requirement for units involves an a priori rejection of the possibility that a trait experiences natural selection and in doing so leads to different replication or reproduction of entities at the lower level. That is, fusing together interactor and replicator/reproducer/reconstitutor requirements, which is not biologically plausible.

An additional advantage of his framework, Godfrey-Smith contends, is that it allows explaining how transitions in individuality work, by explaining how some higher levels get *Darwinized* – or even *de-Darwinized* if the transition is reversed – while lower levels get *de-Darwinized*. The main idea to understanding these processes in Godfrey-Smith's framework is to understand how heritability gets transferred between levels. While conceiving the ETI project as concerning how heritability gets transferred between levels is a correct characterization of the ETI project, it constitutes an incorrect reduction of the units of selection *debates*, which encompass several research questions, to *the singular debate* framed under the restricted set of research questions deriving from the adaptationist ETI project, which is simply one among the many projects in today's biology that addresses questions about the units of selection.

Moreover, Godfrey-Smith's concept of Darwinian individuality aims to reduce questions about the units into a single question where one single object simultaneously fulfils the interactor and reproducer's role, as some critics have noticed (Sterelny 2011; Lloyd 2023), while it is in most cases also a manifestor/type-1 agent (Lloyd 2023), at least with respect to its reproduction regime. This is because the three possible mechanisms that guarantee the existence of collective reproduction in Godfrey-Smith's account are allegedly engineering or trans-temporally accumulated adaptations that have evolved in certain lineages and allow the evolution of further engineering adaptations at that level. Godfrey-Smith's analysis hence confounds the concepts of interactor, manifestor/type-1 agent, and reproducer into a single functional role. This moves him to conclude that there is only one question about units.

We could express Godfrey-Smith's question in our own terms by saying that his one-and-only question about units concerns discovering entities that are simultaneously interactors and reproducers, where their joint reproduction has been achieved by a process of accumulation of adaptations with respect to the sequestration of reproduction at their level.

Philosophy of Biology

The main underlying assumption in Godfrey-Smith's conception of the units is that appealing to different types of objects with distinct functions multiplies entities beyond necessity (Misconception 5, Table 2), and in a manner that would be conceptually inadequate to understand the ETI, where only one role would be necessary (Misconception 4, Table 1).

This analysis conceiving the ETI as a project that requires finding an object that simultaneously fulfils all the functional roles we have isolated in our analysis of units is, again, not necessarily an incorrect analysis of how the ETI project works. But it strikes us as mistaken that Godfrey-Smith discards the DP on these grounds. Asking different questions in combination and requiring that one single object has all the properties required for an affirmative answer to these questions does not mean that the questions are not conceptually different. We contend they are, insofar as they respond to different types of functional roles that may appear independently in different objects. Furthermore, the ETI project can be completely reconstructed by relying on the distinctions made under the DP if one acknowledges that, first, some researchers are interested in *combinations of these questions*, and, second, this combination requires gathering different types of evidence that cannot always be collected together, nor does it always appear together at the same focal level (Table 1; Section 3.3).

Thus, what we consider would be a legitimate conclusion from Godfrey-Smith's analysis in that some research – namely, many ETI questions – on the units of selection focuses on combinations of questions. But what Godfrey-Smith concludes from his analysis is rather more ambitious – a complete denial of the polysemic character of the expression 'unit of selection'. To quote:

> Questions about the "unit" of selection *are not ambiguous*; the units in a selection process are just the entities that make up a Darwinian population at that level. It is always possible to ask a further question: what is the mechanism of inheritance? But that is an optional question about how the patterns of inheritance that give rise to a Darwinian process at a given level come about. (Godfrey-Smith 2009, p. 111, emphasis added)

The essential problem with this type of claim is that it bluntly silences the historical achievements that the DP has played in clarifying the debates about species selection, group selection, and others (Section 2), in favour of an eliminative version of the UP. However, in a charitable interpretation of the eliminative version of the UP defended by Godfrey-Smith, one could argue that his attitude could constitute a faithful analysis of today's biological research. To establish this claim, though, he would need to show that every participant in the debates about the units *now* actually shares the same concept, that is the concept of his Darwinian individual, which simultaneously fuses the concepts of interactor, manifestor/type-1 agent, and reproducer previously isolated in the literature. Perhaps the disagreement about the meaning of the expression 'units of selection' was a topic of the past, something that was salient in the last decades of the twentieth century/early 2000s, but whose period of validity has expired. It is true that this is a possibility, because scientific communities eventually reach agreements, and certain research projects/questions simply fade away. Godfrey-Smith does not pursue this type of analysis himself, but we think Okasha (2006) does (Misconception 5, Table 2). He asserts:

> This [the idea that interactor and reproducer questions need to be fused into a single framework that explains ETI, while assuming other applications of the tripartite framework can be abandoned] is an important consideration, whose implications extend beyond the question of the suitability of the Dawkins–Hull framework. It highlights a subtle transformation in the levels-of-selection question since the discussions of the 1960s and 1970s .... These early discussions tended to take the existence of the biological [reproductive] hierarchy for granted .... But of course the biological hierarchy is itself the product of evolution .... So ideally, *we would like an evolutionary theory which explains how the biological hierarchy came into existence, rather than treating it as a given.* From this perspective, the levels-of-selection question is not simply about identifying the hierarchical levels(s) at which selection *now* acts, which is how it was traditionally conceived, but about identifying the mechanisms which led the various [reproductive] hierarchical levels to evolve in the first place. *Increasingly, evolutionary theorists have turned their attention to this latter question.* (Okasha 2006, p. 16, emphasis added)[33]

Okasha is probably correct in claiming that we need an evolutionary theory that explains how the reproductive hierarchy came into existence, for example as tackled by Maynard Smith and Szathmáry 1995, and Griesemer 2000c (but see fn. 26 concerning the historical origins of the issue). He is also correct that some early bipartite versions of the DP proved inadequate to frame this specific

---

[33] Note that Okasha's claim about the necessity of an evolutionary theory that explains the [reproductive] hierarchy and his conviction that functional analyses cannot do so, while Lewontin's analysis can, may be contested. Of course, if the functionalist version assumes the existence of highly evolved properties, Okasha is correct that it is impossible. But as Griesemer notes, this is not necessary if the functionalist framework only requires relations of compositionality between levels, regardless of the material ways in which this compositionality is realized; in other words, if the project is conceived in metaphysical or ontological terms, without specifying the mechanism, rather than in biological terms, which require its specification. In fact, Griesemer claims that this would be a correct characterization of Lloyd's analysis, and the reason why it would be preferable: 'Lloyd's formal approach only requires a relation of compositionality between entities at different levels for her formal individuation of entities to work. It is therefore important to distinguish between "the biological hierarchy" as employed by biologists, which requires ontological commitment to a structure with empirical content, and the abstract "conceptual" hierarchy defined by Lloyd's formal theory' (Griesemer 2005, p. 71; see also Section 2.1.1)

problem, especially for the role attributed to the replicator (Misconception 3, Table 2), and for the conflation of interactor questions with manifestor/type-1 agent questions (Misconception 1, Table 2). But assuming that the origin of the reproductive hierarchy is *the only or even the primary problem* that evolutionary theory addresses is a mistake, which extends to the conclusions of their analysis of the urgency posed by the ETI project way beyond the limits of what the project itself establishes.

Secondly, the claim that an increasing number of evolutionary theorists have turned their attention to the questions about the origin of the reproductive hierarchy, which requires the combination of interactor, reproducer, and manifestor/type-1 agent questions (a combination that as we have explained is possible even if one follows the DP, see Misconception 4, Table 2), even if true, does not invalidate the original project of disambiguation. This is especially so if at least some researchers *now* still need to frame their projects in different terms by appealing to different meanings of 'units of selection', requiring different types of evidence, and using different types of modelling (Table 1). Failing to appreciate that this is a necessity for today's biology leads to – or is based upon – a misconception of the plurality of debates about units.

4.3 Reinventing the Disambiguating Project in Recent Biology and Philosophy of Biology

As a matter of fact, and supporting our claims, even some authors explicitly endorsing the notion of Darwinian individual and the univocal approach to the debates about units it entails have recently taken a step back, resurrecting the DP. A contemporary debate where this tendency can be perceived concerns the recent disputes about the role of holobionts (host-microbiome multispecies consortia) as units of selection (see Suárez 2018; Suárez & Stencel 2020, for a review; Lloyd 2018). A good illustration of the resurrection of the DP is provided by the work of Stencel and Wloch-Salamon (2018, p. 201), who rely on Queller and Strassman's (2009, 2016) adaptationist conception of the 'organism' – to be read as manifestor/type-1 agent – to claim that questions about reproduction or the reproducer (a key ingredient in the concept of *Darwinian individuality*) should be dissociated from questions about manifestors/type-1 agents.[34]

Stencel & Wloch-Salamon claim:

> This way of conceptualising organisms [manifestors/type-1 agents] makes a great deal of sense from an evolutionary perspective, because organisms need to deal with certain environmental obstacles in every generation. For them

---

[34] Technically speaking, Stencel and Wloch-Salamon's *organism* is a mixture of the *interactor* and the *manifestor/type-1 agent*.

to do so, the elements from which they are built must function in a co-ordinated way to perform certain tasks, such as development, growth, and the digestion of a particular resource, in order to survive and reproduce. [i.e., they must be manifestors/type-1 agents]. Thus, seeing an organism as a system built of elements that co-operate to maintain its structure makes sense, because evolution is all about making such systems much more co-operative in order to perform tasks 'assigned' by the environment.

The most interesting aspect of this conceptualisation of the organism [manifestor] is that it is indifferent to the mode of inheritance of the interacting elements [replicator/reproducer/reconstitutor]. Therefore, since the emphasis is placed on co-operation among entities, it embraces elements that are not necessarily co-inherited over generations, such as genes inside a nuclear genome, but, as in the interactions presented above, are characterised by a high level of co-operation and a low level of conflict. Indeed, two units may have evolved a high level of interdependence, sophisticated mechanisms of communication, and even some degree of integration of their biochemical machinery [thus being manifestors/type-1 agents], but still might reproduce independently [thus not being replicators/reproducers]

Thus, their analysis requires both manifestors/type-1 agents and reproducers in order to work in evolution, which suggests that the DP is still necessary to address different live evolutionary questions.

A similar recent tendency to resurrect the DP project by those previously following the Darwinian individuals approach can be found in Doolittle and Booth (2017, pp. 18–19), who in contrast with Stencel and Wloch-Salamon believe that the key distinction to be resurrected is that between interactors and replicators. To quote:

> We do not in general favor the replicator/interactor model (Hull 1980) as the best or only way to understand selection, but it is worth noting that it can be coherently employed in this context, and in a way that is substantially different from other proposals. Lloyd [2018] for example, conceives holobionts as interactors that promote the differential success of the lineages of cells that make them up. We offer an alternative in which holobionts are seen as interactors that promote the differential success of the interaction patterns they instantiate. The replicators in our model are abstract functional relationships, not cell lineages.

And, more recently, in Lean et al. (2022), they claim to be linking community genetics with the (*sic*) replicator–interactor theory to understand how communities can evolve by natural selection. More surprisingly, they even claim that if one wants to understand how communities evolve by natural selection, one must conceive them as interactors housing replicators/reproducers/reconstitutors at the lower level, rather than as a single level that both interacts *and* reproduces. While this may seem surprising to those holding an adaptationist perspective, it seems to us an obvious reinforcement of the necessity of the concepts distinguished under the tripartite framework, and also of the necessity of combining adaptationist modelling with the models from those working on the evolutionary change school of genetics (e.g., Lloyd and Wade 2019).

Overall, this shows that, while the ETI project – and the types of concepts that were put forward to capture the type of questions about the units that this project assumes – has required expanding the concept of replicator to the more inclusive concept of the *reproducer* as a key notion within the DP, it by no means entails that the rest of questions and meanings previously isolated under the DP project have lost their relevance.

On the contrary, the ETI project simply reinvigorates the relevance of the DP, and the necessity of keeping these meanings distinct even while some research projects may address some of the questions isolated by these meanings *simultaneously* or in combination. It would be a mistake, as we have demonstrated, to assume that because the DP distinguishes meanings of the expression 'units of selection', types of questions addressed by relying on these different meanings, and types of evidence and modelling necessary to answer these questions affirmatively, that these questions cannot be addressed simultaneously when the research project requires it (Misconception 5, Table 2; Table 1). The fact that two concepts are distinct does not entail that they are not related, or that one object cannot simultaneously satisfy both concepts, or even that there is a nomological connection between the different concepts in some cases. Failing to appreciate this is a metaphysically inadequate picture of the theory, context, and evidence, and has induced a serious epistemological and methodological confusion in research about units, both in the past and in the present.

We thus contend that the necessity of these three roles is particularly clear in *today's* biology, as they have been reintroduced again by several other researchers using a different nomenclature for each of them (Gardner & Grafen 2009; Gardner & Welch 2011; Bijma 2014; Gardner 2015; Goodnight 2015; Doolittle & Booth 2017; Okasha 2018; Stencel & Proszewska 2018; Bourrat 2022; Lean et al. 2022), but with the goal of clarifying similar debates about the units to the ones Lloyd aimed at clarifying when she introduced them.

4.4 A Further Ado to the Tripartite Framework: The Reconstitutor

More recently, several authors have called attention to specific processes in which certain phenotypic traits seem to reappear consistently across generations without replication or reproducing processes at that specific level. Thus, these cases seem to be violating some of the conditions required by the concept of the replicator and the reproducer as we have introduced and advocated for them in this work.

One of these cases involves some phenotypic states caused by small RNAs. For instance, in a case documented by Rechavi et al. (2014) and analyzed by Veigl (2017), the authors showed how starvation in *Caenorhabditis elegans* caused changes in the proportions in the pool of small RNAs in the cells. Not surprisingly, these changes were inherited if the environmental trigger (starvation) continued but, surprisingly, they were also preserved for a few generations *even in the absence of the environmental trigger*. Note that in these cases there is no transgenerational material overlap, because even if the parent can transmit the small RNA to its offspring, it cannot transmit the changes in the *proportions* in the pool of small RNAs responsible for the phenotype in question. This is so both for cases of sexual reproduction and hermaphrodite self-fertilization. Additionally, the parent cannot formally cause the reappearance of the trait, as it has no potential to generate the environmental trigger causing it in the first place, and it apparently has no other genetic causation through its process of sexual reproduction. Therefore, a trait is transgenerationally preserved for a brief span of generations, yet there are not replicators or reproducers at the level to support the phenotype or interactor.

Holobionts, on the other hand, can constitute another good example of the transgenerational preservation of phenotypic traits in the absence of replication or reproduction at the focal level of holobiont interactor selection. An example of this occurs in the heredity of sanguivory (obligate blood diet) in vampire bats, as studied by Mendoza et al. (2018) and analyzed by Suárez (2020) and Suárez and Triviño (2020).

In their study, Mendoza et al. showed that vampire bats are not adapted to cope with sanguivory in the absence of their microbiome (their complements of bacteria, virus, and fungal microbes living throughout their intestinal tract), given that the traits encoded in their genome are insufficient to cope with the extremely demanding challenges of a strictly blood-based diet. Furthermore, they also showed that many of the traits subject to selection in the microbiome of vampire bats are directly connected to sanguivory, and cannot be detected in the free-living counterparts of the microorganisms inhabiting the microbiome. Additionally, parents do not transmit the microbiome to their offspring vertically, through their eggs, or reproductive processes: this rather needs to be reassembled in every new generation. The reassembly requires adjusting the selection pressures over thousands or millions of bacterial species, each of which has an independent reproductive pattern and can also inhabit different environments. Yet, the reassembly reconstitutes a vampire bat which is apt to cope with the environmental challenges of a sanguivory diet. However, the units recreating the holobiont come from many different sources and escape parental control. Thus, sanguivory is transgenerationally recreated in vampire bat holobionts, but there are no replicator or reproducers at the holobiont level.[35]

Drawing on this, in a recent work, Veigl et al. (2022) introduce the concept of the *reconstitutor* to account for these cases. In their view, a *reconstitutor* is an entity that results from the transgenerational active relationship between independently reproducing elements that join together in the next generation to recreate specific phenotypic variants that appeared in previous generations.

An important characteristic of reconstitutors is that they do not in principle need any causal influence from the previous generation of reconstitutors, including material overlap and/or any type of formal influence from the parents. Instead, they would simply re-assemble and recreate the network of interactions giving rise to the functioning phenotype. Importantly, the concept of the reconstitutor does not deny that there may be replicators and/or reproducers at some levels in the system. In fact, small RNAs are replicated, and the host and its microbial taxa are reproduced. Their point is rather that these processes do not occur *at the focal level where trait preservation is occurring*, and thus the necessity of expanding the replicator/reproducer concepts to a more extensive one, covering also these examples. Drawing on this, Veigl et al. (2022) argue that the reconstitutor must be an essential unit for thinking of some cases of the evolutionary process. We think it captures some nuances of the causal role originally attributed to the replicator but not captured either by the replicator or the reproducer, given the specific material properties attributed to both entities (see Veigl et al. 2022).

Note that the reconstitutor can be mapped onto the tripartite framework we have advocated in this work as a role to capture hereditary relationships over time. Therefore, it is a concept that expands on the replicator and reproducer but fulfils the same functional role that these concepts fulfil. Therefore, we propose to adopt the reconstitutor in our analysis as a unit that biologists sometimes inquire about when doing research about units.

Conclusion

In this Element, we have provided a methodological defence of what we have called the DP (disambiguating project) about the units of selection, according to which the expression 'unit of selection' is polysemic and refers to different types of functional roles as captured by different types of modelling practices in

---

[35] A similar reassembly from outside sources is also found when the human infant is born, in the founding of their intestinal microbiome from their environment (see Chiu and Gilbert 2015).

today's biology. These were most noticeably disputes between the evolutionary change and the adaptationist schools in evolution – developed to answer specific non-equivalent but sometimes related questions (Table 1). More specifically, we have distinguished three meanings that should not be conflated in debates about units: interactor, replicator/reproducer/reconstitutor, and manifestor/type-1 agent.

We have provided solid evidence that even while some contemporary biological research encompasses all these questions, which are asked simultaneously or in combination, this is not the case in all biological research about units. Hence, the different meanings isolated under the DP (and possibly more to be discovered) should be highlighted in the analysis of the units of selection to avoid unnecessary confusions, and to prevent biologists (or philosophers) from talking past each other. This Element has been built as a systematic response to the contemporary tendency that, building on research on the ETI, may lead to the embrace of an eliminativist version of the UP (unitary project), which rejects the ambiguity of the expression 'units of selection'. We have shown this is incorrect, and it is not a profitable research avenue to do research about units.

Our argument has combined systematic analysis of the different projects about units with a historical perspective on the development of these very projects, particularly the DP, and the ETI project. While systematicity plays a fundamental role in properly addressing philosophical questions and understanding the diverging meanings of the expression 'units of selection', the historical perspective is the only one providing the ultimate rationale of why the DP was proposed in the first place and why the different meanings (interactor, replicator/reproducer/reconstitutor, and manifestor/type-1 agent) were isolated and added in the past (Section 2). It also provides the rationale of why research on the ETI has moved some authors (most prominently, Griesemer 2000a, 2000b, 2005) to further expand the DP to include a more generalized and expanded version of the replicator – the reproducer – while others have decided to renounce the DP and argue that the expression unit of selection is not ambiguous (Section 4).

We have provided evidence to show that the DP, in its tripartite formulation, is still necessary in contemporary biological research, as some biologists, as well as philosophers, still demand to disambiguate meanings *that had been disambiguated in the past*. The evidence we have offered includes recent works by Goodnight (2015), Bijma (2014), Okasha (2018), Stencel and Wloch-Salamon (2018), Doolittle and Booth (2017), and others. Table 3 summarizes the main sources of evidence about the reintroduction of the distinctions that we have identified and summarized across the Element.

Table 3 Reintroduction of the three different meanings of 'unit of selection', during the last decade

Authors	Distinction they recognize	Nomenclature they use	Where we cite the evidence
Gardner & Welch 2011	Interactors and manifestor/type-1 agent	Target of selection (= interactor) Maximizing agent (= manifestor/type-1 agent)	Section 3
Bijma 2014	Interactors and replicators/reproducers/reconstitutors	IGEs on trait-values models (= interactor) Fitness-centred models (= replicator)	Section 3
Okasha 2018	Interactors and manifestor/type-1 agent	MLS modelling (= interactor) Type-1 agent (=Manifestor/type-1 agent)	Section 3
Stencel & Wloch-Salamon 2018	manifestor/type-1 agent and replicators/reproducers/reconstitutors	Units of cooperation (= manifestor/type-1 agent) Reproductive mode (= replicator/reproducer/reconstitutor)	Section 4
Doolittle & Booth 2017	Interactors and replicators/reproducers/reconstitutors	Interactors Replicators	Section 4
Lean et al. (2022)	Interactors, replicators/reproducers/reconstitutors and manifestor/type-1 agent	Interactors Replicators	Section 4

These authors propose similar distinctions to the ones previously advocated under the tripartite framework of the DP to articulate with precision certain contemporary debates in the biological sciences. We have analyzed how the distinctions made in these works relate to three concepts originally isolated by Lloyd (1988/1994, 1992, 2001, 2023) and have contended that the fact that contemporary biologists still require these distinctions poses a strong argument in favour of keeping the DP.

Additionally, we have provided evidence to show that most criticisms of the DP are partial and based on several misconceptions about the rationale of the project and about the multiplicity of concepts ultimately isolated and their real meaning (conceived as the type of evidence required to show an appropriate application of the concept to a specific biological phenomenon). Table 2 summarizes these misconceptions and Table 1 spells out succinctly the different meanings of the expression 'units of selection' that have been isolated, and what these meanings have been designed to capture. We conclude this Element by contending that there is not one question about units, but several clearly distinguishable questions, *each requiring a distinct concept of 'unit'*. Hence the expression 'units of selection' is polysemic. Keeping different meanings distinct is fundamental to avoid confusion in contemporary biology. The DP is here to stay, and the reader is invited to investigate whether some contemporary biological practices require the introduction of new meanings.

Glossary

Adaptationist School of Evolution/Methodological Adaptationism:
School of evolutionary thought that seeks to understand the origin of 'engineering' adaptations in certain biological lineages. Researchers working under this tradition focus their investigation on some products of the evolutionary process, and build kin selection/inclusive fitness models and optimality models that account for these products as designed or optimal traits to perform functions in the biological lineage.

Evolutionary Change School of Evolution: School of evolutionary thought that seeks to understand the processes of evolution and selection themselves, irrespectively of the outcomes they may give rise to (i.e., irrespectively of whether they result in engineering adaptations). Researchers working under this tradition build multilevel selection (MLS) models to determine the selection coefficient at one specific level, and predict how evolution will occur in the population in virtue of such coefficients.

Disambiguating Project (DP): General name that we use to refer to the family of conceptions about the units of selection according to which the concept is polysemic, and may legitimately refer to different non-co-extensional concepts that need to be carefully distinguished from one another. Under this project, it is assumed that part of the disputes about the units of selection result from the use of different meanings under the same generic label 'unit of selection'.

Unitary Project (UP): General name that we use to refer to the family of conceptions about the units of selection according to which the concept requires philosophical clarification because it is hard to define, but not because it is ambiguous or polysemic; on the contrary, it is assumed that the expression has a univocal meaning, even while this needs to be investigated. Under this view, it is assumed that disputes about the units result from incorrect uses of the expression, or incorrect application of the concept to empirical cases. Defenders of the UP hold that a unit of selection is an entity in a population of entities that simultaneously express heritable phenotypic variance in fitness at the focal level.

Tinkering/Engineering or Trans-temporally Accumulated Adaptation:
Trait that serves a 'designed' purpose for its bearer and does so because it results from a history of selection and accumulation at that level. Engineering adaptations have the 'right type' of causal history underlying their evolutionary origin. Determining that a trait is an engineering adaptation requires a

selection model that shows that it is possible to produce such an outcome at the focal level. They must be distinguished from product-of-selection adaptations, or group benefits.

Group Benefit/Product-of-Selection Adaptation: Trait that results from a selection process acting on a population that increases the fitness of its bearers now, but does not need to be the result of an accumulation process of selection at the level. We call a trait a 'product of selection adaptation' when it results from higher-level selection (i.e., non-zero selection coefficient at the higher level) that produces a higher-level benefit by chance. Alternatively, a group benefit can also be a non-adaptive trait, in case it results from 'lucky' selection at the lower level. Group benefits, especially of the product-of-selection type, must be distinguished from engineering or trans-temporally accumulated adaptations.

(Interactor/Replicator) Bipartite Framework/Interpretation: Framework for conceiving the units of selection that distinguishes two meanings: replicator and interactor. It is an early version of the DP.

Tripartite Framework/Interpretation: Framework for conceiving the units of selection that distinguishes three meanings: replicator/reproducer/reconstitutor, interactor, and manifestor of adaptation/type-1 agent. It is our preferred version of the DP.

Recipe Approach: Framework for conceiving the units of selection as the entities that bear a certain closed set of properties, similar to a closed recipe. Different authors defend different 'recipes', depending on the properties they consider necessary for an entity to be unit of selection. It is the standard form of UP about the units of selection.

Evolutionary Transitions in Individuality or Major Transitions in Individuality (ETI): Project in evolutionary biology that investigates how different modes of reproduction evolved, and how what were once independently reproducing units became integrated into a single unit of reproduction. In this work, we are mostly concerned with adaptationist versions of the ETI.

References

Amundson, Ron. 2001. 'Adaptation, Development, and the Quest for Common Ground'. In: Orzack, SH and Sober, E (eds.) *Adaptationism and Optimality.* Cambridge University Press, New York, pp. 303–34.

Arnold, Stevan J and Michael J Wade. 1984a. 'On the Measurement of Natural and Sexual Selection: Applications'. *Evolution* 38 (4): 720–34.

1984b. 'On the Measurement of Natural and Sexual Selection: Theory'. *Evolution* 38 (4): 709–19.

Biernaskie, Jay M and Kevin R Foster. 2016. 'Ecology and Multilevel Selection Explain Aggression in Spider Colonies'. *Ecology Letters* 19 (8): 873–9.

Bijma, Piter. 2014. 'The Quantitative Genetics of Indirect Genetic Effects: A Selective Review of Modelling Issues'. *Heredity* 112 (1): 61–69.

Bijma, Piter and Michael J Wade. 2008. 'The Joint Effects of Kin, Multilevel Selection and Indirect Genetic Effects on Response to Genetic Selection'. *Journal of Evolutionary Biology* 21: 1175–88.

Bijma, Piter, William M Muir, Esther D Ellen, Jason B Wolf, and Johan AM Van Arendonk. 2007a. 'Multilevel Selection 2: Estimating the Genetic Parameters Determining Inheritance and Response to Selection'. *Genetics* 175 (1): 289–99.

Bijma, Piter, William M Muir, and Johan AM Van Arendonk. 2007b. 'Multilevel Selection 1: Quantitative Genetics of Inheritance and Response to Selection'. *Genetics* 175 (1): 277–88.

Bock, Walter. 1980. 'The Definition and Recognition of Biological Adaptation'. *Integrative and Comparative Biology* 20(1): 217–27.

Booth, Austin. 2014. 'Symbiosis, Selection, and Individuality'. *Biology & Philosophy* 29 (5): 657–73.

Borrello, Mark E. 2010. *Evolutionary Restraints: The Contentious History of Group Selection.* University of Chicago Press, Chicago.

Bourrat, Pierrick. 2002. 'A New Set of Criteria for Units of Selection'. *Biol Theory* 17: 263–75.

Bourrat, Pierrick. 2019. 'Evolutionary Transitions in Heritability and Individuality'. *Theory in Biosciences* 138 (2): 305–23.

2021. *Facts, Conventions, and the Levels of Selection.* Elements in the Philosophy of Biology. Cambridge University Press, Cambridge.

Bourrat, Pierrick and Paul E Griffiths. 2018. 'Multispecies Individuals'. *History and Philosophy of the Life Sciences* 40 (2): 1–23.

Brandon, Robert. 1981. 'Biological Teleology: Questions and Explanations'. *Studies in History and Philosophy of Science Part A* 12 (2): 91–105.

———. 1982. 'The Levels of Selection'. *Proceedings of the Philosophy of Science Association* 1982: 315–23.

———. 1985. 'Adaptation Explanations: Are Adaptations for the Good of Replicators or Interactors?' In: David J. Depew, Bruce H. Weber (eds.) *Evolution at a Crossroads: The New Biology and the New Philosophy of Science*. MIT Press, Cambridge, MA, pp. 81–96.

———. 1988. 'The Levels of Selection: A Hierarchy of Interactors'. In: H.C. Plotkin (ed.) *The Role of Behavior in Evolution*. MIT Press, Cambridge, MA, pp. 51–71.

———. 1990. *Adaptation and Environment*. Vol. 1040. Princeton University Press, Princeton, NJ.

Burian, Richard D. 1992. 'Adaptation: Historical Perspectives'. In: Keller, EF and Lloyd, EA (eds.) *Keywords in Evolutionary Biology*. Harvard University Press, Cambridge, MA.

Buss, Leo W. 1987. *The Evolution of Individuality*. Princeton University Press, Princeton, NJ.

Carroll, Sean B. 2005. *Endless Forms Most Beautiful: The New Science of Evo Devo and the Making of the Animal Kingdom*. Norton, New York.

Charnov, Eric. 1982. *The Theory of Sex Allocation*. Princeton University Press, Princeton.

Chiu, Lynn, Gilbert, Scott F. 2015. 'The Birth of the Holobiont: Multi-species Birthing through Mutual Scaffolding and Niche Construction'. *Biosemiotics* 8: 191–210.

Clarke, Ellen. 2013. 'The Multiple Realizability of Biological Individuals'. *The Journal of Philosophy* 110 (8): 413–35.

———. 2014. 'Origins of Evolutionary Transitions'. *Journal of Biosciences* 39 (2): 303–17.

Damuth, John and Lorraine Heisler. 1988. 'Alternative Formulations of Multilevel Selection'. *Biology and Philosophy* 3 (4): 407–30.

Dawkins, Richard. 1982a. 'Replicators and Vehicles'. *Current Problems in Sociobiology* 45 (64): 45–64.

———. 1982b. *The Extended Phenotype*. Vol. 8. Oxford University Press, Oxford.

———. 1976/2016. *The Selfish Gene*. Oxford University Press, Oxford, UK.

Díaz, Javier Suárez. 2017. 'El Mecanismo Evolutivo de Margulis y Los Niveles de Selección'. *Contrastes: Revista Internacional de Filosofía* 20 (1): 7–24.

Doolittle, W Ford, and Eric Bapteste. 2007. 'Pattern Pluralism and the Tree of Life Hypothesis'. *Proceedings of the National Academy of Sciences* 104 (7): 2043–9.

Doolittle, Ford and Austin Booth. 2017. 'It's the Song, Not the Singer: An Exploration of Holobiosis and Evolutionary Theory'. *Biology & Philosophy* 32(1): 5–24.

Dugatkin, Lee A and Hudson K Reeve. 1994. 'Behavioral Ecology and Levels of Selection: Dissolving the Group Selection Controversy'. *Advances in the Study of Behavior* 23: 101–33.

Dupré, John and Maureen A O'Malley. 2009. 'Varieties of Living Things: Life at the Intersection of Lineage and Metabolism'. *Philosophy, Theory, and Practice in Biology* 1 (20130604): 1–25.

Fontana, Walter and Leo W Buss. 1994. '"The Arrival of the Fittest": Toward a Theory of Biological Organization'. *Bulletin of Mathematical Biology* 56 (1): 1–64.

Forsdyke, Donald R. 2010. 'George Romanes, William Bateson, and Darwin's "Weak Point"'. *Notes and Records of the Royal Society* 64 (2): 139–54.

Gardner, Andy. 2009. 'Adaptation as Organism Design'. *Biol. Lett.* 5: 861–4.

Gardner, Andy. 2015. 'Group Selection versus Group Adaptation'. *Nature* 524 (7566): E3–4.

Gardner, Andy and Alan Grafen. 2009. 'Capturing the Superorganism: A Formal Theory of Group Adaptation'. *Journal of Evolutionary Biology* 22 (4): 659–71.

Gardner, Andy and John Welch. 2011. 'A Formal Theory of the Selfish Gene'. *Journal of Evolutionary Biology* 24(8): 1801–13.

Gardner, Andy, Stuart A. West, and Geoff Wild. 2011. 'The Genetical Theory of Kin Selection'. *Journal of Evolutionary Biology* 24: 1020–43.

Godfrey-Smith, Peter. 2001. 'Three Kinds of Adaptationism'. In: Orzack, SH, and Sober, E (eds.) *Adaptationism and Optimality*. Cambridge University Press, Cambridge, pp. 344–62.

2007. 'Conditions for Evolution by Natural Selection'. *The Journal of Philosophy* 104 (10): 489–516.

2009. *Darwinian Populations and Natural Selection*. Oxford University Press, Oxford, UK.

2011. 'Agents and Acacias: Replies to Dennett, Sterelny, and Queller'. *Biology & Philosophy* 26(4): 501–15.

2013. 'Darwinian Individuals'. *From Groups to Individuals: Evolution and Emerging Individuality* 16: 17–36.

2015. 'Reproduction, Symbiosis, and the Eukaryotic Cell'. *Proceedings of the National Academy of Sciences* 112 (33): 10120–5.

Goodnight, Charles J. 2015. 'Multilevel Selection Theory and Evidence: A Critique of Gardner, 2015'. *Journal of Evolutionary Biology* 28 (9): 1734–46.

Goodnight, Charles J and Lori Stevens. 1997. 'Experimental Studies of Group Selection: What Do They Tell Us about Group Selection in Nature?' *The American Naturalist* 150 (S1): s59–79.

Goodnight Charles J, Schwartz James M, Stevens Lori. (1992). 'Contextual Analysis of Models of Group Selection, Soft Selection, Hard Selection and the Evolution of Altruism'. *Am. Nat.* 140 (5): 743–61.

Gould, Stephen Jay. 1977. 'Caring Groups and Selfish Genes'. *Natural History* 86 (10): 20–24.

2002. *The Structure of Evolutionary Theory*. Harvard University Press, Cambridge MA.

Gould, Stephen Jay and Elisabeth Lloyd. 1999. 'Individuality and Adaptation across Levels of Selection: How Shall We Name and Generalize the Unit of Darwinism?' *Proceedings of the National Academy of Sciences* 96 (21): 11904–9.

Gould, Stephen Jay and Richard C Lewontin. 1979. 'The Spandrels of San Marco and the Panglossian Paradigm'. *Proceedings of the Royal Society* 205: 581–98.

Grant, B Rosemary and Peter R Grant. 1989. *Evolutionary Dynamics of a Natural Population: The Large Cactus Finch of the Galápagos*. University of Chicago Press, Chicago, IL.

Grant, Peter R and Bara Rosemary Grant. 2020. *How and Why Species Multiply*. Princeton University Press, Princeton, NJ.

Grantham, Todd A. 1994. 'Putting the Cart Back behind the Horse: Group Selection Does Not Require That Groups Be "Organisms"'. *Behavioral and Brain Sciences* 17 (4): 622–3.

Green, Sara. 2014. 'A Philosophical Evaluation of Adaptationism as a Heuristic Strategy'. *Acta Biotheor* 62(4): 479–98.

Griesemer, James R. 2000a. 'Development, Culture, and the Units of Inheritance'. *Philosophy of Science* 67: S348–68.

2000b. 'Reproduction and the Reduction of Genetics'. In: Peter J. Beurton, Raphael Falk, Hans-Jörg Rheinberger (eds.) *The Concept of the Gene in Development and Evolution: Historical and Epistemological Perspectives*. Cambridge University Press, Cambridge, UK, pp. 240–85.

2000c. 'The Units of Evolutionary Transition'. *Selection* 1 (1–3): 67–80.

2003. 'The Philosophical Significance of Gánti's Work'. In: Gánti, T (ed.) *The Principles of Life*. Oxford University Press, New York, pp. 169–85.

2005. 'The Informational Gene and the Substantial Body: On the Generalization of Evolutionary Theory by Abstraction'. *Idealization XII: Correcting the Model. Idealization and Abstraction in the Sciences* 86: 59–115.

Griesemer, James R and Michael J Wade. 2000. 'Populational Heritability: Extending Punnett Square Concepts to Evolution at the Metapopulation Level'. *Biology and Philosophy* 15 (1): 1–17.

Haldane, John Burdon. 1932/1990. *The Causes of Evolution*. Vol. 5. Princeton University Press, Princeton, NJ.

Hamilton, William D. 1975. Innate Social Aptitudes in Man: An Approach from Evolutionary Genetics. In: Robin Fox (ed.) *Biosocial Anthropology*. New York, Wiley, pp. 133–55.

Heisler, I Lorraine and John Damuth. 1987. 'A Method for Analyzing Selection in Hierarchically Structured Populations'. *The American Naturalist* 130 (4): 582–602.

Holsinger, Kent E. 1994. 'Groups as Vehicles and Replicators: The Problem of Group-Level Adaptation'. *Behavioral and Brain Sciences* 17 (4): 626–7.

Hull, David L. 1980. 'Individuality and Selection'. *Annual Review of Ecology and Systematics* 11: 311–32.

1988a. *Science as a Process*. University of Chicago Press, Chicago, IL.

1988b. 'Interactors versus Vehicles'. In: Henry C. Plotkin (ed.) *The Role of Behavior in Evolution*. MIT Press, Cambridge, MA, pp. 19–50.

2001. *Science and Selection: Essays on Biological Evolution and the Philosophy of Science*. Cambridge University Press, Cambridge, UK.

Jablonski, David. 2008. 'Species Selection: Theory and Data'. *Annual Review of Ecology, Evolution, and Systematics* 39: 501–24.

Jablonski, David and Gene Hunt. 2006. 'Larval Ecology, Geographic Range, and Species Survivorship in Cretaceous Mollusks: Organismic Versus Species-Level Explanations'. *American Naturalist* 168 (4): 556–64.

Jeler, Ciprian. 2020. 'Explanatory Goals and Explanatory Means in Multilevel Selection Theory'. *History and Philosophy of the Life Sciences* 42 (3): 1–24.

Keller, Evelyn F and Elisabeth A. Lloyd. 1992. *Keywords in Evolutionary Biology*. Harvard University Press, Cambridge, MA.

Lande, Russell and Stevan J Arnold. 1983. 'The Measurement of Selection on Correlated Characters'. *Evolution* 37: 1210–26.

Lean, Christopher H, Doolittle, Ford W and Joseph P Bielawski. 2022. 'Community-level Evolutionary Processes: Linking Community Genetics with Replicator-Interactor Theory'. *PNAS* 119 (46): e2202538119.

Levins, Richard and Richard Lewontin. 1985. *The Dialectical Biologist*. Harvard University Press, Cambridge, MA.

Lewens, Tim. 2009. 'Seven Types of Adaptationism'. *Biol Philos* 24(2): 161–82.

Lewontin, Richard C. 1962. 'Interdeme Selection Controlling a Polymorphism in the House Mouse'. *The American Naturalist* 96 (887): 65–78.

1970. 'The Units of Selection'. *Annual Review of Ecology and Systematics*. 1: 1–18.

1978. 'Adaptation'. *Scientific American* 239 (3): 212–30.

Lewontin, Richard C and Leslie Clarence Dunn. 1960. 'The Evolutionary Dynamics of a Polymorphism in the House Mouse'. *Genetics* 45 (6): 705–22.

Lloyd, Elisabeth A. 1986. 'Evaluation of Evidence in Group Selection Debates'. *Proceedings of the Philosophy of Science Association* 1986: 483–93.

1992. 'Unit of Selection'. In: Fox Keller, E and Lloyd, EA (eds.) *Keywords in Evolutionary Biology*. Harvard University Press, Cambridge, MA, pp. 334–40.

1988/1994. *The Structure and Confirmation of Evolutionary Theory*. Princeton University Press, Princeton, NJ .

1994. 'Rx: Distinguish Group Selection from Group Adaptation'. *Behavioral and Brain Sciences* 17 (4): 628–9.

1999. 'Altruism Revisited. Review of *Unto Others* by Elliott Sober and David Sloan Wilson'. *Quarterly Review of Biology* 74: 447–9.

2001. 'An Anatomy of the Units of Selection Debates'. In: Rama S. Singh, Costas B. Krimbas, Diane B. Paul, and John Beatty (eds.) *Thinking about Evolution: Historical, Philosophical, and Political Perspectives* Vol 2. Cambridge University Press, Cambridge, pp. 267–90.

2015. 'Adaptationism and the Logic of Research Questions: How to Think Clearly about Evolutionary Causes'. *Biological Theory* 10(4): 343–62.

2018. 'Holobionts as Units of Selection: Holobionts as Interactors, Reproducers, and Manifestors of Adaptation'. In: Gissis SB, Lamm E, Shavit A (eds.) *Landscapes of Collectivity in the Life Sciences*. MIT Press, London, pp 305–24.

2021. *Adaptation*. Elements in the Philosophy of Biology. Cambridge University Press, Cambridge.

2023. 'Units and Levels of Selection'. In: Zalta, Edward N (ed.) *Stanford Encyclopedia of Philosophy*.

Lloyd, Elisabeth A and Marcus W Feldman. 2002. 'Commentary: Evolutionary Psychology: A View from Evolutionary Biology'. *Psychological Inquiry* 13 (2): 150–6.

Lloyd, Elisabeth A and Stephen J Gould. 1993. 'Species Selection on Variability'. *Proceedings of the National Academy of Sciences* 90 (2): 595–9.

2017. 'Exaptation Revisited: Changes Imposed by Evolutionary Psychologists and Behavioral Biologists'. *Biological Theory* 12: 50–65.

Lloyd, Elisabeth A. 2005. 'Why the Gene Will Not Return'. *Philosophy of Science* 72 (2): 287–310.

Lloyd, Elisabeth A, Richard C Lewontin, and Marcus W Feldman. 2008. 'The Generational Cycle of State Spaces and Adequate Genetical Representation'. *Philosophy of Science* 75 (2): 140–56.

Lloyd, Elisabeth A, and Michael J Wade. 2019. 'Criteria for Holobionts from Community Genetics'. *Biological Theory* 14 (3): 151–70.

Maynard Smith, John. 1964. 'Group Selection and Kin Selection'. *Nature* 201 (4924): 1145–7.

1978. 'Optimization Theory in Evolution'. *Annual Review of Ecology and Systematics* 9: 31–56.

2001. 'Reconciling Marxism and Darwin'. *Evolution* 55 (7): 1496–8.

Maynard Smith, John and Eors Szathmáry. 1995. *The Major Transitions in Evolution*. Oxford University Press, Oxford UK.

Mayr, Ernst. 1983. 'How to Carry Out the Adaptationist Program?' *The American Naturalist* 121: 324–34.

Mendoza, M. Lisandra Zepeda, Zijun Xiong, Marina Escalera-Zamudio, et al. 2018. 'Hologenomic Adaptations Underlying the Evolution of Sanguivory in the Common Vampire Bat'. *Nature Ecology and Evolution* 2: 659–68.

Merlin, Francesca. 2017. 'Limited Extended Inheritance'. In: Philippe Huneman, Denis Walsh (eds.) *Challenges to Evolutionary Theory: Development, Inheritance, Adaptation*. Oxford University Press, Oxford, UK, pp. 263–79.

Merlin, Francesca and Livio Riboli-Sasco. 2017. 'Mapping Biological Transmission: An Empirical, Dynamical, and Evolutionary Approach'. *Acta Biotheoretica* 65 (2): 97–115.

Millstein, Roberta L. 2002. 'Are Random Drift and Natural Selection Conceptually Distinct?' *Biology and Philosophy* 17 (1): 33–53.

Okasha, Samir. 2003. 'Biological Altruism'. In: Zalta, Edward N (ed.) *Stanford Encyclopedia of Philosophy*. https://plato.stanford.edu/archives/sum2020/entries/altruism-biological/.

2006. *Evolution and the Levels of Selection*. Oxford University Press, Oxford.

2010. 'Altruism Researchers Must Cooperate'. *Nature* 467: 653–5.

2016. 'The Relation between Kin and Multilevel Selection: An Approach Using Causal Graphs'. *The British Journal for the Philosophy of Science* 67 (2): 435–70.

2018. *Agents and Goals in Evolution*. Oxford University Press, Oxford.

O'Malley, Maureen A and Russell Powell. 2016. 'Major Problems in Evolutionary Transitions: How a Metabolic Perspective Can Enrich Our Understanding of Macroevolution'. *Biol Philos* 31: 159–89.

Orzack, Steven H and Elliot Sober. 1994. 'Optimality Models and the Test of Adaptationism'. *The American Naturalist* 143: 361–80.

Orzack, Steven Hecht and Patrick Forber. 2017. *Adaptationism*. In: Zalta, Edward N (ed.) *Stanford Encyclopedia of Philosophy*. https://plato.stanford.edu/archives/spr2017/entries/adaptationism/.

Papale, François. 2020. 'Evolution by Means of Natural Selection without Reproduction: Revamping Lewontin's Account'. *Synthese* 198: 10429–10455.

Pigliucci, Massimo and Jonathan Kaplan. 2000. 'The Rise and Fall of Dr. Pangloss: Adaptationism and the Spandrels Paper 20 Years Later'. *Trends in Ecology & Evolution* 15(2): 66–70.

Queller, David C and Joan E Strassmann. 2009. 'Beyond Society: The Evolution of Organismality'. *Philosophical Transactions of the Royal Society B: Biological Sciences* 364 (1533): 3143–55.

2016. 'Problems of Multi-Species Organisms: Endosymbionts to Holobionts'. *Biology & Philosophy* 31 (6): 855–73.

Rechavi, Oded, Leah Houri-Ze'evi, Sarit Anava, et al. 2014. 'Starvation-induced Transgenerational Inheritance of Small RNAs in *C. elegans*'. *Cell* 158(2): 277–87.

Reeve, Hudson Kern and Laurent Keller. 1999. 'Levels of Selection: Burying the Units of Selection Debate and Unearthing the Crucial New Issues'. In Keller, L (ed.) *Levels of selection in evolution*. Princeton University Press, Princeton, pp. 3–14.

Reeve, Hudson Kern and Paul W Sherman. 1993. 'Adaptation and the Goals of Evolutionary Research'. *The Quarterly Review of Biology* 68(1): 1–32.

Salmon, Wesley C. 1971. *Statistical Explanation and Statistical Relevance*. Vol. 69. University of Pittsburgh Press, Pittsburgh, Pennsylvania, PA.

Sansom, Roger. 2003. 'Constraining the Adaptationism Debate'. *Biology and Philosophy* 18: 493–512.

Skillings, Derek. 2016. 'Holobionts and the Ecology of Organisms: Multi-Species Communities or Integrated Individuals?' *Biology & Philosophy* 31 (6): 875–92.

Smith, Eric Alden. 1994. 'Semantics, Theory, and Methodological Individualism in the Group-Selection Controversy'. *Behavioral and Brain Sciences* 17 (4): 636–7.

Sober, Elliott. 1984. *The Nature of Selection: Evolutionary Theory in Philosophical Focus*. Vol. 95. University of Chicago Press, Chicago, IL.

Sober, Elliott and David Sloan Wilson. 1993. *Philosophy of Biology*. Vol. 45. Westview Press, Boulder, Colorado, CO.

1998. *Unto Others: The Evolution and Psychology of Unselfish Behavior*. Harvard University Press, Cambridge, MA.

Stencel, Adrian. 2016. 'The Relativity of Darwinian Populations and the Ecology of Endosymbiosis'. *Biology & Philosophy* 31 (5): 619–37.

Stencel, Adrian and Agnieszka M Proszewska. 2018. 'How Research on Microbiomes Is Changing Biology: A Discussion on the Concept of the Organism'. *Foundations of Science* 23 (4): 603–20.

Stencel, Adrian and Dominika M. Wloch-Salamon. 2018. 'Some Theoretical Insights into the Hologenome Theory of Evolution and the Role of Microbes in Speciation'. *Theory in Biosciences* 137(2): 197–206.

Sterelny, Kim. 2011. 'Darwinian Spaces: Peter Godfrey-Smith on Selection and Evolution'. *Biology & Philosophy* 26 (4): 489–500.

Suárez, Javier. 2018. '"The Importance of Symbiosis in Philosophy of Biology: An Analysis of the Current Debate on Biological Individuality and Its Historical Roots"'. *Symbiosis* 76 (2): 77–96.

2019. 'The Hologenome Concept of Evolution: A Philosophical and Biological Study'. PhD Dissertation. University of Exeter, Exeter.

2020. The Stability of Traits Conception of the Hologenome: An Evolutionary Account of Holobiont Individuality. *HPLS* 42 (11) (2020).

2021. 'El Holobionte/Hologenoma Como Nivel de Selección: Una Aproximación a La Evolución de Los Consorcios de Múltiples Especies'. *THEORIA. An International Journal for Theory, History and Foundations of Science* 36 (1): 81–112.

Suárez, Javier and Adrian Stencel. 2020. 'A Part-Dependent Account of Biological Individuality: Why Holobionts Are Individuals and Ecosystems Simultaneously'. *Biological Reviews* 95 (5): 1308–24.

Suárez, Javier and Vanessa Triviño. 2020. 'What Is a Hologenomic Adaptation? Emergent Individuality and Inter-Identity in Multispecies Systems'. *Frontiers in Psychology* 11: 1–15.

Szathmáry, Eörs. 2015. 'Toward Major Evolutionary Transitions Theory 2.0'. *Proceedings of the National Academy of Sciences of the USA* 112: 10104–11.

Veigl, Sophie J 2017. 'Use/Disuse Paradigms Are Ubiquitous Concepts in Characterizing the Process of Inheritance'. *RNA Biology* 14(12): 1700–04.

Veigl, Sophie J, Suárez Javier, and Adrian Stencel. 2022. 'Rethinking Hereditary Relations: The Reconstitutor as the Evolutionary Unit of Heredity'. *Synthese* 200: 1–42.

Wade, Michael J. 1977. 'An Experimental Study of Group Selection'. *Evolution* 31 (1): 134–53.

1979. 'Sexual Selection and Variance in Reproductive Success'. *The American Naturalist* 114 (5): 742–7.

1980a. 'Kin Selection: Its Components'. *Science* 210(4470): 665–7.

1980b. 'An Experimental Study of Kin Selection'. *Evolution* 34(5): 844–55.

1985. 'Soft Selection, Hard Selection, Kin Selection, and Group Selection'. *The American Naturalist* 125 (1): 61–73.

2016. *Adaptation in Metapopulations: How Interaction Changes Evolution.* University of Chicago Press, Chicago, IL.

Wade, Michael J and David E McCauley. 1980. 'Group Selection: The Phenotypic and Genotypic Differentiation of Small Populations'. *Evolution* 34 (4): 799–812.

1984. 'Group Selection: The Interaction of Local Deme Size and Migration in the Differentiation of Small Populations'. *Evolution* 38 (5): 1047–58.

1988. 'Extinction and Recolonization: Their Effects on the Genetic Differentiation of Local Populations'. *Evolution* 42 (5): 995–1005.

Wade, Michael J and James Griesemer. 1988. 'Population Heritability: Empirical Studies of Evolution in Metapopulations'. *The American Naturalist* 151(2): 135–47.

Wagner, Günter P, Chi-hua Chiu, and Manfred Laubichler. 2000. 'Developmental Evolution as a Mechanistic Science: The Inference from Developmental Mechanisms to Evolutionary Processes'. *American Zoologist* 40: 819–31.

Williams, George Christopher. 1966. *Adaptation and Natural Selection: A Critique of Some Current Evolutionary Thought.* Vol. 75. Princeton University Press, Princeton, NJ.

1990. 'Review of *The Structure and Confirmation of Evolutionary Theory* by Elisabeth Lloyd'. *Quarterly Review of Biology* 65: 504.

1992. *Natural Selection: Domains, Levels, and Challenges.* Oxford University Press, Oxford.

Wilson, David Sloan. 1975. 'A Theory of Group Selection'. *Proceedings of the National Academy of Sciences* 72 (1): 143–6.

Wilson, David Sloan and Elliot Sober. 1994. 'Reintroducing Group Selection to the Human Behavioral Sciences'. *Behavioral and Brain Sciences* 17(4): 585–608.

Wimsatt, William C. 1980a. 'Reductionistic Research Strategies and Their Biases in the Units of Selection Controversy'. In: Thomas Nickles (ed.) *Scientific Discovery: Case Studies.Reidel, Dordrecht,* pp. 213–59.

1980b. 'The Units of Selection and the Structure of the Multi-Level Genome'. *Proceedings of the Philosophy of Science Association* 1980 (2): 122–83.

Wynne-Edwards, Vero Copner. 1962. *Animal Dispersion: In Relation to Social Behaviour.* Oliver and Boyd, Edinburgh.

Acknowledgements

Javier Suárez thanks the Narodowe Centrum Nauki (Grant Opus No: 2019/35/B/HS1/01998) for financial support.

Cambridge Elements ≡

Philosophy of Biology

Grant Ramsey

KU Leuven, Belgium

Grant Ramsey is a BOFZAP research professor at the Institute of Philosophy, KU Leuven, Belgium. His work centers on philosophical problems at the foundation of evolutionary biology. He has been awarded the Popper Prize twice for his work in this area. He also publishes in the philosophy of animal behavior, human nature and the moral emotions. He runs the Ramsey Lab (theramseylab.org), a highly collaborative research group focused on issues in the philosophy of the life sciences.

Michael Ruse

Florida State University

Michael Ruse is the Lucyle T. Werkmeister Professor of Philosophy and the Director of the Program in the History and Philosophy of Science at Florida State University. He is Professor Emeritus at the University of Guelph, in Ontario, Canada. He is a former Guggenheim fellow and Gifford lecturer. He is the author or editor of over sixty books, most recently *Darwinism as Religion: What Literature Tells Us about Evolution; On Purpose; The Problem of War: Darwinism, Christianity, and their Battle to Understand Human Conflict;* and *A Meaning to Life.*

About the Series

This Cambridge Elements series provides concise and structured introductions to all of the central topics in the philosophy of biology. Contributors to the series are cutting-edge researchers who offer balanced, comprehensive coverage of multiple perspectives, while also developing new ideas and arguments from a unique viewpoint.

Cambridge Elements ≡

Philosophy of Biology

Elements in the Series

Stem Cells
Melinda Bonnie Fagan

The Metaphysics of Biology
John Dupré

Facts, Conventions, and the Levels of Selection
Pierrick Bourrat

The Causal Structure of Natural Selection
Charles H. Pence

Philosophy of Developmental Biology
Marcel Weber

Evolution, Morality and the Fabric of Society
R. Paul Thompson

Structure and Function
Rose Novick

Hylomorphism
William M. R. Simpson

Biological Individuality
Alison K. McConwell

Human Nature
Grant Ramsey

Ecological Complexity
Alkistis Elliott-Graves

Units of Selection
Javier Suárez and Elisabeth A. Lloyd

A full series listing is available at www.cambridge.org/EPBY

Printed in the United States
by Baker & Taylor Publisher Services